新楼盘 44
NEWHOUSE 图解地产与设计

年度最佳楼盘与景观

中国林业出版社

上海中建设计

恪守诚信原则
聚焦客户需求
打造精品设计
呈现优质服务

上海中建建筑设计院有限公司
SHANGHAI ZHONGJIAN ARCHITECTURAL DESIGN INSTITUTE CO., LTD.

上海中建·上海总部
浦东新区东方路989号
中 达 广 场 1 2 楼
电话：86 - 21 - 6875 8810

上海中建·西安分院
南 二 环 西 段 202号
九 座 花 园 611室
电话：86 - 29 - 8837 8506

HUAYI
DESIGN
華藝設計

建 筑　　规 划　　室 内　　景 观
Architectural　Planning　Interior　Landscape

总部地址：深圳市福田区华富路航都大厦14、15层
邮编：518031
总机：(86-755)8379 0262
传真：(86-755)8379 0289
E-mail：master@huayidesign.com
www.huayidesign.com

香港　深圳　北京　上海　南京　武汉　重庆　广州　厦门　成都

LUCAS 奥德景观
DESIGN GROUP

西安阳光城-西西安小镇

深圳市奥德景观规划设计有限公司

 简介

公司坐落于著名的蛇口湾畔，深圳最有影响力的创意设计基地：南海意库；
公司前身为深圳市卢卡斯景观设计有限公司，是由 2003 年成立于香港的卢卡斯联盟（香港）国际设计有限公司在世界设计之都：深圳设立的中国境内唯一公司。
于 2012 年 1 月获得中华人民共和国国家旅游局正式认定：旅游规划设计乙级设计资质。

 公司专注于

居住区景观与规划设计（含旅游地产）
商业综合体景观与规划设计（含购物公园、写字楼及创意园区）
城市规划及空间设计
市民公园设计
酒店与渡假村景观规划与设计
旅游策划及规划设计

 公司目标

倚重当下的中国的文化渊源结合世界潮流，尊重地域情感，在中国打造具强烈地域特征的、风格化的、国际化的，具前瞻性、可再生的的城市景观、人居环境、风情渡假区及自然保护区。

地　　址：深圳市南山区蛇口兴华路南海意库 2 栋 410 室
电　　话：0755-86270761
传　　真：0755-86270762
邮　　箱：lucasgroup_lucas@163.com
网　　址：www.lucas-designgroup.com

lindun jianzhujingguan

Residential real estate
Commercial real estate
Tourism real estate
Resort Hotel
Municipal planning
Campus planning
Park planning
Regional planning

住宅地産 | 商業地産 | 旅游地産 | 度假酒店 | 市政規劃 | 校園規劃 | 公園規劃 | 區域規劃

L&D 靈頓景觀 (USA)
建築景觀
法國建築師協會會員單位

美國靈頓建築景觀設計有限公司是專業從事城市規劃和城市設計、風景區與公園景觀規劃、風景旅游渡假景觀、酒店環境景觀、高尚住宅小區景觀及工程設計的國際知名公司。深圳靈頓建築景觀設計有限公司是其專爲中國地區設置的設計顧問公司，公司以國外多位設計師爲公司主幹，美國公司作爲主要規劃整體和宏觀控制，以共同協作，共同致力于同一項目上，使設計工程在整體規劃及局部處理上都得到精心設計。公司以全新的理念引導市場，以專業的服務介入市場，以全程的服務方式開拓市場，使公司得到穩步和迅速的發展，業務範圍不斷擴大。尤其在高尚住宅小區景觀設計中取的優异的成績，其設計工程多次獲國家和地區的獎勵。

http://www.szld2005.com   中國 深圳 福田區 紅荔路花卉世界313號

TEL: (0755) 8621 0770   FAX: (0755) 8621 0772   P.C: 518000   EMAIL: szld2000@163.com   179049195@QQ.com

美國靈頓建築景觀設計有限公司   深圳靈頓建築景觀設計有限公司   雲南靈頓園林綠化工程有限公司

哲思国际

哲思国际是较早进入中国市场并立足的有国际影响力的澳大利亚设计公司。以华南珠三角起步，所主持与参与的项目已扩展到长江三角洲、大西南和华北地区，已成为中国市场一支活跃的设计力量。

规划与设计项目范围包括：

> 区域性规划
> 大型社区规划
> 商业、零售及配套
> 居住区开发
> 大型公建设施
> 展览及会议中心
> 办公楼及配套
> 酒店及度假村设施
> 文化教育设施
> 工业及科技园区

澳洲总部：
ZENX INTERNATIONAL PTY. LTD.

电话：61-7-3701 2882
传真：61-7-3701 2929
邮箱：info@zenx.com.au

中国广州：
哲思（广州）建筑设计咨询有限公司

电话：020-3882 3593
传真：020-3882 3598
邮箱：zenx_gz@vip.163.com

中国深圳：
深圳市哲思建筑设计咨询有限公司

电话：0755-8344 7500
传真：0755-8344 7503
邮箱：zenx_sz@vip.163.com

建筑设计　总体规划　都市设计　景观建筑设计

www.zenx.com.au

长春天茂庄园示范区荣耀呈现

GVL怡境景观
GREENVIEW LANDSCAPE DESIGN LIMITED

- 景观设计 — **L**andscape Design
- 旅游度假项目规划 — **R**esorts and Leisure Planning
- 市政项目规划 — **U**rban Planning Design
- 居住环境项目规划 — **C**ommunity Planning
- 公园及娱乐项目规划 — **P**arks and Entertainment Planning and Design

英国国家园景
工业协会海外会员
British Association of Landscape Industries, Overseas Full Member of BALI

美国景观设计师协会企业会员
Corporate Member Of American Society Of Landscape Architects

广 州 总 部
地 址：珠江新城华夏路49号津滨腾越大厦南塔8楼
邮 编：510623
电 话：(020) 87690558   87695498
              38032762   38032729
传 真：(020) 87697706
邮 箱：greenv@163.com
网 址：www.greenview.com.cn

香 港
地 址：北角渣华道18号嘉汇商业大厦2106室
邮箱编号：070432
传 真：(852) 22934388

北 京
地 址：朝阳区亚运村阳光广场B2-1701室
邮 编：100101
传 真：(010) 64977992
电 话：(010) 64975897

www.greenview.com.cn

@GVL怡境景观
http://weibo.com/gvlcn

www.greenview.com.cn

前言 EDITOR'S NOTE

成长 收获 展望
GROWTH, HARVEST AND VISTA

转眼2012接近尾声,又到一年年末时。这一年是我们杂志革新的一年、成长的一年、收获的一年。中英双语版本、与时代紧密结合的栏目专题、专家理论观点与设计解读、最新最具创意的设计作品等等,让全新版本的面世得到了广大新老读者的认可和喜爱,对此我们倍感欣慰。也非常感谢广大的合作客户和读者一直以来对我们的支持与鼓励,新的一年,期待能为大家带来更多惊喜。

回望2012,类型丰富、数量庞大的楼盘及景观系列设计作品层出不穷,不仅展示出各大设计公司、事务所等不断提升的设计实力,也展现出设计方面的各类新技术、新理念等。年末盘点之际,本期杂志精选出部分年度最佳楼盘与景观设计作品作为专题,与您一同深入品鉴和探讨。

展望未来,我们愿以更新颖的思维、更严谨的态度、更踏实的行动为您呈现更好更精彩的读品刊物。

In a twinkle of an eye, it's approaching the end of 2012. In this year, New House has experienced reform, growth and harvest. We are delighted that the Chinese-English bilingual version, features closely bond to times, expert's viewpoint and design literacy and the latest and most creative works, etc have helped us win the approval and fondness from both old and new readers. We would like to express our appreciation for the support and encourage on our clients and readers side. In the coming new year, we will commit ourselves to bring more surprise to our readers.

Retrospecting into 2012, abundant types and large number of new houses have emerged endlessly, demonstrating the continual improvement of design companies and new techniques and new ideas coming up in design world. In this last inventory-check magazine of 2012, we selected from the best houses and landscape as feature to probe and review in depth with our readers.

Looking into the future, we would present our readers with better publications in the spirit of more innovative thinking, more rigorous attitude and more earnest action.

jiatu@foxmail.com

NEWHOUSE 图解地产与设计

2012年 总第44期

面向全国上万家地产商决策层、设计院、建筑商、材料商、专业服务商的精准发行

指导单位 INSTRUCTION UNIT
亚太地产研究中心

出品人 PUBLISHER
杨小燕 YANG XIAOYAN

主编 CHIEF EDITOR
王 志 WANG ZHI

副主编 ASSOCIATE EDITOR
熊 冕 XIONG MIAN

编辑记者 EDITOR REPOTERS
唐秋琳 TANG QIULIN
钟梅英 ZHONG MEIYING
胡明俊 HU MINGJUN
康小平 KANG XIAOPING
吴 辉 WU HUI
曾伊莎 ZENG YISHA
曹丹莉 CAO DANLI
朱秋敏 ZHU QIUMIN
王盼青 WANG PANQING

设计总监 ART DIRECTORS
杨先周 YANG XIANZHOU
何其梅 HE QIMEI

美术编辑 ART EDITOR
詹婷婷 ZHAN TINGTING

国内推广 DOMESTIC PROMOTION
广州佳图文化传播有限公司

市场总监 MARKET MANAGER
周中一 ZHOU ZHONGYI

市场部 MARKETING DEPARTMENT
方立平 FANG LIPING
熊 光 XIONG GUANG
王 迎 WANG YING
杨先凤 YANG XIANFENG
熊 灿 XIONG CAN
刘 佳 LIU JIA

图书仕版编目（CIP）数据

新楼盘. 年度最佳楼盘与景观：汉英对照 / 佳图文化主编.
—— 北京：中国林业出版社，2012.12
ISBN 978-7-5038-6857-3

Ⅰ.①新... Ⅱ.①佳... Ⅲ.①建筑设计 - 中国 - 现代 - 图集 Ⅳ.①TU206

中国版本图书馆CIP数据核字(2012)第030658号
出版：中国林业出版社
主编：佳图文化
责任编辑：李顺 许琳
印刷：利丰雅高印刷(深圳)有限公司

特邀顾问专家 SPECIAL EXPERTS (排名不分先后)

赵红红 ZHAO HONGHONG	赵士超 ZHAO SHICHAO
王向荣 WANG XIANGRONG	孙 虎 SUN HU
陈世民 CHEN SHIMIN	梅卫平 MEI WEIPING
陈跃中 CHEN YUEZHONG	林世彤 LIN SHITONG
邓 明 DENG MING	熊 冕 XIONG MIAN
冼剑雄 XIAN JIANXIONG	周 原 ZHOU YUAN
陈宏良 CHEN HONGLIANG	李焯忠 LI ZHUOZHONG
胡海波 HU HAIBO	原帅让 YUAN SHUAIRANG
程大鹏 CHENG DAPENG	王 颖 WANG YING
范 强 FAN QIANG	周 敏 ZHOU MIN
白祖华 BAI ZUHUA	王志强 WANG ZHIQIANG / DAVID BEDJAI
杨承刚 YANG CHENGGANG	陈英梅 CHEN YINGMEI
黄宇奘 HUANG YUZANG	吴应忠 WU YINGZHONG
梅 坚 MEI JIAN	曾繁柏 ZENG FANBO
陈 亮 CHEN LIANG	朱黎青 ZHU LIQING
张 朴 ZHANG PU	曹一勇 CAO YIYONG
盛宇宏 SHENG YUHONG	冀 峰 JI FENG
范文峰 FAN WENFENG	滕赛岚 TENG SAILAN
彭 涛 PENG TAO	王 毅 WANG YI
徐农思 XU NONGSI	陆 强 LU QIANG
田 兵 TIAN BING	徐 峰 XU FENG
曾卫东 ZENG WEIDONG	张奕和 EDWARD Y. ZHANG
马素明 MA SUMING	郑竞晖 ZHENG JINGHUI
仇益国 QIU YIGUO	刘海东 LIU HAIDONG
李宝章 LI BAOZHANG	凌 敏 LING MIN
李方悦 LI FANGYUE	谢锐何 XIE RUIHE
林 毅 LIN YI	陈 锐 CHEN RUI
陈 航 CHEN HANG	姜 圣 JIANG SHENG
范 勇 FAN YONG	

编辑部地址：广州市海珠区新港西路3号银华大厦4楼
电话：020—89090386/42/49、28905912
传真：020—89091650

北京办：王府井大街277号好友写字楼2416
电话：010—65266908　**传真**：010—65266908

深圳办：深圳市福田区彩田路彩福大厦B座23F
电话：0755—83592526　**传真**：0755—83592536

协办单位 CO—ORGANIZER

广州市金冕建筑设计有限公司　熊冕 总设计师
地址：广州市天河区珠江西路5号国际金融中心主塔21楼06—08单元
TEL：020—88832190　88832191
http://www.kingmade.com

AECF 上海颐朗建筑设计咨询有限公司　巴学天 上海区总经理
地址：上海市杨浦区大连路970号1308室
TEL：021—65909515　　FAX：021—65909526
http://www.yl—aecf.com

WEBSITE COOPERATION MEDIA
网站合作媒体

SouFun 搜房网

副理事长单位 DEPUTY CHAIRMAN

华森设计 HSArchitects
华森建筑与工程设计顾问有限公司　邓明　广州公司总经理
地址：深圳市南山区滨海之窗办公楼6层
　　　广州市越秀区德政北路538号达үan大厦26楼
TEL：0755—86126888　020—83276688
http://www.huasen.com.cn　E—mail:hsgzaa@21cn.net

广州瀚华建筑设计有限公司　冼剑雄　董事长
地址：广州市天河区黄埔大道中311号羊城创意产业园2—21栋
TEL：020—38031268　FAX：020—38031269
http://www.hanhua.cn
E—mail：hanhua—design@21cn.net

CSCEC
上海中建建筑设计院有限公司　徐峰　董事长
地址：上海市浦东新区东方路989号中达广场12楼
TEL：021—68758810　FAX：021—68758813
http://www.shzjy.com
E—mail：csaa@shzjy.com

常务理事单位 EXECUTIVE DIRECTOR OF UNIT

POF 华域普风
深圳市华域普风设计有限公司　梅坚　执行董事
地址：深圳市南山区海德三道海岸城华座1306—1310
TEL：0755—86290985　FAX：0755—86290409
http://www.pofart.com

WDCE 华通国际
华通设计顾问工程有限公司
地址：北京市西城区西直门南小街135号西派国际C—Park3号楼
TEL：8610—83957395　8610—83957390
http://www.wdce.com.cn

TEAMER ARCH
天萌（中国）建筑设计机构　陈宏良　总建筑师
地址：广州市天河区员村四横路128号红专厂F9栋天萌建筑馆
TEL：020—37857429　FAX：020—37857590
http://www.teamer—arch.com

GVL 怡境景观
GVL国际怡境景观设计有限公司　彭涛　中国区董事及设计总监
地址：广州市珠江新城华夏路49号津滨腾越大厦南塔8楼
TEL：020—87690558　FAX：020—87697706
http://www.greenview.com.cn

TENIO
天友建筑设计股份有限公司　马素明　总建筑师
地址：北京市海淀区西四环北路158号慧科大厦7F（方案中心）
TEL：010—88592005　FAX：010—88229435
http://www.tenio.com

R-land 源树景观
R—LAND 北京源树景观规划设计事务所　白祖华　所长
地址：北京朝阳区朝外大街怡景园5—9B
TEL：010—85626992/3　FAX：010—85625520
http://www.ys—chn.com

L&A 奥雅
奥雅设计集团　李宝章　首席设计师
深圳总部地址：深圳蛇口南海意库5栋302
TEL：0755—26826690　FAX：0755—26826694
http://www.aoya—hk.com

北京寰亚国际建筑设计有限公司　赵士超　董事长
地址：北京市朝阳区琨莎中心1号楼1701
TEL：010—65797775　FAX：010—84682075
http://www.hygjjz.com

山水比德
广州山水比德景观设计有限公司　孙虎　董事总经理兼首席设计师
地址：广州市天河区珠江新城临江大道685号红专厂F19
TEL：020—37039822/823/825　FAX：020—37039770
http://www.gz—spi.com

OCEANICA
奥森国际景观规划设计有限公司　李焯忠　董事长
地址：深圳市南山区南海大道1061号喜士登大厦四楼
TEL：0755—26828246　86275795　FAX：0755—26822543
http://www.oc—la.com

SJS 四季园林
广州市四季园林设计工程有限公司　原帅让　总经理兼设计总监
地址：广州市天河区龙怡路117号银汇大厦2505
TEL：020—38273170　FAX：020—86682658
http://www.gz—siji.com

深圳市雅蓝图景观工程有限公司　周敏　设计董事
地址：深圳市南山区南海大道2009号新能源大厦A座6D
TEL：0755—26650631/26650632　FAX：0755—26650623
http://www.yalantu.com

BM 佰邦建筑
深圳市佰邦建筑顾问有限公司　迟春儒　总经理
地址：深圳市南山区兴工路8号美年广场1栋804
TEL：0755—86229594　FAX：0755—86229559
http://www.pba—arch.com

NEW ERA 新纪元设计
北京新纪元建筑工程设计有限公司　曾繁柏　董事长
地址：北京市海淀区小马厂6号华天大厦20层
TEL：010—63483388　FAX：010—63265003
http://www.bjxinjiyuan.com

博地澜屋建筑设计
北京博地澜屋建筑规划设计有限公司　曹一勇　总设计师
地址：北京市海淀区中关村南大街31号神舟大厦8层
TEL：010—68118690　FAX：010—68118691
http://www.buildinglife.cn

HPA
HPA上海海波建筑设计事务所　陈立波、吴海青　公司合伙人
地址：上海中山西路1279弄6号楼国峰科技大厦11层
TEL：021—51168290　FAX：021—51168240
http://www.hpa.cn

HUAYI 華藝設計
香港华艺设计顾问（深圳）有限公司　林毅　总建筑师
地址：深圳市福田区华富路航都大厦14、15楼
TEL：0755—83790262　FAX：0755—83790289
http://www.huayidesign.com

哲思（广州）建筑设计咨询有限公司　郑竟晖　总经理
地址：广州市天河区天河北路626号保利中宇广场A栋1001
TEL：020—38823593　FAX：020—38823598
http://www.zenx.com.au

理事单位 COUNCIL MEMBERS （排名不分先后）

BAC 柏澳景觀
广州市柏澳景观有限公司　徐农思　总经理
地址：广州市天河区广园东路2191号时代新世界中心南塔2704室
TEL：020—87569202　FAX：020—87635002
http://www.bacdesign.com.cn

CREO Architects&Engineers
中房集团建筑设计有限公司　范强　总经理/总建筑师
地址：北京市海淀区百万庄建设部院内
TEL：010—68347818

BHAD
北京奥思得建筑设计有限公司　杨承冈　董事总经理
地址：北京朝阳区东三环中路39号建外SOHO16号楼2903~2905
TEL：86—10—58692509/19/39　58692523

CA CHEN SHIMIN 陈世民
陈世民建筑师事务所有限公司　陈世民　董事长
地址：深圳市福田中心区益田路4068号卓越时代广场4楼
TEL：0755—88262516/429

JACC 嘉柯园林
广州嘉柯林景观设计有限公司　陈航　执行董事
地址：广州市珠江新城华夏路49号津滨腾越大厦北塔506—507座
TEL：020—38032521/23　FAX：020—38032679
http://www.jacc—hk.com

JNC 侨恩国际
侨恩国际（美国）建筑设计咨询有限公司
地址：重庆市渝北区龙湖MOCO4栋20—5
TEL：023—88197325　FAX：023—88197323
http://www.jnc—china.com

CONCORD 西迪国际
CDG国际设计机构　林世彤　董事长
地址：北京海淀区长春桥11号万柳亿城中心A座10/13层
TEL：010—58815603　58815633　FAX：010—58815637
http://www.cdgcanada.com

广州市圆美环境艺术设计有限公司　陈英梅　设计总监
地址：广州市海珠区宝岗大道杏坛大街56号二层之五
TEL：020—34267226　83628481　FAX：020—34267226
http://www.gzyuanmei.com

WEME 唯美景观-中国
上海唯美景观设计工程有限公司　朱黎青　董事、总经理
地址：上海市徐虹中路20号2—202室
TEL：021—61122209　FAX：021—61139033
http://www.wemechina.com

OC
上海金创源建筑设计事务所有限公司　王毅　总建筑师
地址：上海杨浦区黄兴路1858号701—703室
TEL：021—55062106　FAX：021—55062106—807
http://www.odci.com.cn

L&D 建筑景观
深圳灵顿建筑景观设计有限公司　刘海东　董事长
地址：深圳福田区红荔路花卉世界313号
TEL：0755—86210770　FAX：0755—86210772
http://www.szld2005.com

LUCAS DESIGN GROUP 奥德景观
深圳市奥德景观规划设计有限公司　凌敏　董事总经理、首席设计师
地址：深圳市南山区蛇口海上世界南海意库2栋410#
TEL：0755—86270761　FAX：0755—86270762
http://www.lucas—designgroup.com

广州邦景园林绿化设计有限公司　谢锐何　董事及设计总监
地址：广州市天河北路175号祥龙花园晖祥阁2504/05
TEL：020—87510037　FAX：020—38468069
http://www.bonjinglandscape.com

上海天隐建筑设计有限公司　陈航　执行董事
地址：上海市杨浦区国康路100号国际设计中心1402室
TEL：021-65988000　FAX：021—65982798
http://www.skyarchdesign.com

目录 CONTENTS

011　前言 EDITOR'S NOTE

016　资讯 INFORMATION

名家名盘 MASTER AND MASTERPIECE

020　成都中德英伦联邦：典雅高贵的英伦建筑 风情浪漫的园林景观
ELEGANT BRITISH-STYLE ARCHITECTURE AND ROMANTIC GARDEN LANDSCAPE

专访 INTERVIEW

028　建筑的美学价值、时代特征与可持续发展
——访北京奥思得建筑设计有限公司董事总经理 杨承冈
THE AESTHETIC VALUE, CHARACTERISTIC OF EPOCH AND SUSTAINABLE DEVELOPMENT OF ARCHITECTURE

032　营造是设计与施工之间的有效推进剂
——访广州市圆美环境艺术设计有限公司技术总监 吴应忠
CREATION IS AN EFFECTIVE PROPELLANT BETWEEN DESIGN AND CONSTRUCTION

036　十年磨一剑：特色高端景观设计的实践者
——访上海唯美景观设计工程有限公司
A DECADE OF PRACTICE FOR CHARACTERISTIC AND HIGH-END LANDSCAPE DESIGN

新景观 NEW LANDSCAPE

040　扬州·中集紫金文昌景观设计：
古典优雅 精致从容的新古典主义人文景观
ELEGANT AND SOPHISTICATED NEOCLASSICAL CULTURAL LANDSCAPE

046　阳江淡墨幽居：清新淡雅、意境优美的养生庭院景观
FRESH AND PASTEL LIFE-PRESERVING GARDEN LANDSCAPE WITH ELEGANT REALM

专题 FEATURE

054　浅谈楼盘景观主题化设计要点
WITH CONCISE REMARKS ON LANDSCAPE THEME DESIGN FOR HOUSING PROJECTS

056	鹭湖：师法自然、低调奢华的精致别墅社区 IMITATING NATURE, LOW-PROFILE LUXURIOUS EXQUISITE VILLA COMMUNITY
064	南昌天沐•君湖：新亚洲风情休闲社区 NEW ASIAN-INSPIRED LEISURE COMMUNITY
072	南京（马群）紫园：紫气东来 时尚尊贵 AUSPICIOUSNESS, FASHION AND DIGNITY
078	时代南沙山湖海一期： 融合"山、湖、海"元素的简约现代居住空间 INTEGRATE THE ELEMENTS OF "MOUNTAIN, LAKE, SEA" WITH CONCISE MODERN RESIDENTIAL SPACE
084	佛山和丰颖苑：中西合璧、古韵有致的多元化景观空间 ANCIENT DIVERSIFIED LANDSCAPE SPACE COMBINING CHINESE AND WESTERN ELEMENTS

064

新特色 NEW CHARACTERISTICS

090　五矿•哈施塔特别墅区：奥地利风情演绎的山水交响曲
　　　MOUNTAINS AND WATERS SYMPHONY OF AUSTRIAN EXPRESSION

新空间 NEW SPACE

098　深圳联路芷峪澜湾花园B1-A户型样板房：简洁舒适 精致优雅
　　　SIMPLE AND COMFORTABLE, EXQUISITE AND ELEGANT

新创意 NEW IDEA

102　圣•科洛马•德•格拉曼历镇滨江住宅项目二期：
　　　多重高度 极富韵律感滨江住区
　　　MULTIPLE HEIGHT, RHYTHMICAL RIVERFRONT HOUSE

090

商业地产
COMMERCIAL BUILDINGS

112　国家开发银行三亚研究院：临海环山 功能多元的综合性建筑
　　　MULTI-FUNCTIONAL COMPLEX NEARBY MOUNTAINS AND THE SEA

116　深圳星河龙岗COCO PARK：高品质 一站式购物中心
　　　HIGH-QUALITY, ONE-STOP SHOPPING CENTER

120　深圳湾科技生态城B-TEC项目："绿之舞步"都市综合体
　　　"GREENING AS DANCING" URBAN COMPLEX

124　深圳广电集团有线电视枢纽大厦：富于变化 独具时代特征的集合体
　　　A COMPLEX WITH VARIATIONS AND CHARACTERISTICS OF TIMES

128　深圳创维石岩科技园二期：都市型工业综合体
　　　URBAN INDUSTRIAL COMPLEX

132　厦门航空综合开发基地二、三期工程：灵活便利 应"势"生长
　　　FLEXIBLE AND CONVENIENT, DEVELOPING WITH ADVANTAGES

136　贵阳未来方舟C区环球谷超高层综合体：开敞自然 极具吸引力的城市空间
　　　ATTRACTIVE URBAN SPACE: OPEN AND NATURAL

140　南京中建大厦：美学与技术相结合的现代简约总部形象
　　　MODERN MINIMALIST AESTHETICS AND TECHNOLOGY COMBINED HEADQUARTERS

144　单县文化艺术中心：琴台夜月
　　　LYRE PLATFORM AND NIGHT MOON

102

112

INFORMATION 资讯/地产

CIHAF2012专题报道：破局者的盛宴

2012中国住交会将于12月6日~8日在深圳会展中心举行。进入14个年头的CIAHF中国住交会，将在2012年以全新面貌引爆年底的行业狂欢，开启行业寻找真相之旅。

中国住交会以"资源整合者、平台提供者和商务对接者"的面貌，希望为行业参与者建谋势之灯塔、启破局之方舟，深度转型，开拓蓝海，尤其是国际化视野下的全球资源整合，对迷局进行"熊彼特式的破坏"。中国住交会自创立之日就有着"浓得化不开"的英雄情结，在"三名"这台"地产奥斯卡"中走出过许多英雄，成就过很多大佬，这里拥有着指点英雄、激扬豪杰的话语权和评判体系。

全球政治经济正站在一个全新的周期起点上，政经格局一片混沌，尽管陷入迷局的人们感到有点无所适从，但对于破局者而言，却是一场尽情饕餮的盛宴。现在，中国住交会已经为房地产行业充满激情和勇气的破局者们准备了一席丰盛的宴席。2012年12月，中国深圳，饕餮开始！

CIHAF 2012 FEATURE STORY: A FEAST FOR "GAME-CHANGERS"

CIHAF 2012 will take place in Shenzhen Convention & Exhibition Center during Dec. 6th to Dec.8th. The 14th CIAHF, with different appearance, will ignite an industry carnival at the end of 2012 and start a trip of industry truth looking.

CIAHF will turn to be a resources integrator, a connector linking with platform provider and business, who wants to offer guidance for this industry participants, set off the changes, deeply transform and develop a new blue ocean. Especially integrating the global resources under an international outlook, CIHAF will proceed "Schumpeter destruction" to the current puzzle CIAHF has a strong heroic complex since form its foundation. " Three Famous" Award is the Oscar of the real estate, lots of heroes and important persons come from here, possessing a judging system and the speaking right of judging and inspiring heroes.

The global politics and economies are standing on a new staring point. The politics and economies are chaos, even the person in the puzzle cannot figure out. However, as for the "Game-changers", that is a feast. Nowadays, CIHAF has prepared a rich feast for the passionate and brave game-changers in property industry. In 2012, Dec, Shenzhen, China, the feast begins!

捷得建筑师事务所(JERDE) 担纲整体规划设计
大型生态城市综合体开发项目在嘉兴破土动工

10月27日，台商大陆重大投资项目——台昇国际广场在嘉兴破土动工并举行了盛大的奠基仪式。美国捷得建筑师事务所(JERDE)担纲了项目的整体规划设计，通过运用其独特的"场所创造"理念倾力打造以人行为导向的建筑景点，使人们可以在充满绿色的社区环境中自在生活、尽情工作、休憩、聚会。

台昇国际广场的设计旨在创造一个集休闲、娱乐、金融为一体的综合空间，以吸引国内外访客。整个项目占地面积为18万m^2，由五大地块组成，建筑面积为72万m^2。捷得在项目的中心位置规划了一个综合休闲区，以一个大型户外公共广场为中心，连接各个区域，建造一个愉悦融合的聚会场所。整个项目预计于2018年竣工。

LARGE-SCALE AND ECOLOGICAL URBAN COMPLEX PLANNED BY JERDE BROKE GROUND IN JIAXING

On Oct. 27, Samson International Plaza developed by Taiwanese investor broke ground in Jiaxing. JERDE was commissioned to make the master plan for the project. With the unique idea of "Place Creation", it provides a people-oriented green environment for working, recreation and gatherings.

According to the overall planing, Samson International Plaza is envisioned to be an urban complex integrating recreation, entertainment and finance to attract people from all around the world. Featuring a total land area of 180,000 m^2, composed of five plots, it has a total floor area of 720,000 m^2. In the center of the developed, JERDE designed a leisure area around a large open square to connect different plots and provide spaces for pleasant gatherings. The project is expected to be completed by 2018.

限购令悬念倒计时

年底将至，不少城市的房地产限购政策即将到期，海口市4日正式对外宣布，明年将继续执行限购政策，但其他一些曾明确公布限购截止日期的城市至今尚未表态。临近年底，海口市成为全国第一个表态明年限购不放松的城市。

HOUSING RESTRICTION POLICIES SUSPENSE COUNTING DOWN

Getting into this year's end, housing restriction policies of many cities will soon fall due. Haikou city officially declared in 4th that the housing restriction policies in Haikou would continue next year. But other cities that ever clearly publicized the deadline of housing restriction policies haven't declared. By the year's end, Haikou City is the first one that declared next year would not ease restrictions.

广州国际设计周"七待"情现

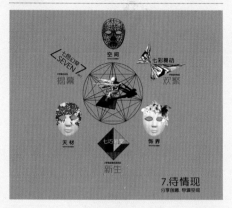

以七年孕育中国设计界梦想与追求的广州国际设计周,将于12月7~9日在广州保利世贸博览馆拉开大幕。记者从组委会获得最新消息,本届广州国际设计周特邀国内知名创意团队——和马建设机构为首席设计顾问,并由其创意总监马劲夫亲自导演空间,打造中国迄今为止最具设计追求与商业价值的大型设计展览。
2012是广州国际设计周7周年生日。马劲夫及其团队由此为切口,呼应本届广州国际设计周"7彩"主题,运用雕塑艺术、装置艺术、装饰艺术等手段,结合绝对设计馆,装饰艺术馆,创新材料馆三大主题展馆特性,进行了大胆而细腻的创意设计,力求呈现出一场异彩纷呈,玲珑多面的设计盛会。

GUANGZHOU DESIGN WEEK: HERE, YOU'RE WELCOME

On December 7-9, Guangzhou Design Week 2012 will be grandly held in Guangzhou Poly World Trade Center. Known from the committee, the famous design group – wellmark Group Office., was commissioned to be the Principal Consultant. And its creative director Ma Jinfu led the team and designed this exhibition to highlight design atmosphere and maximize the commercial value at the same time.
It is the 7th Guangzhou Design Week since its first launch in 2008. Ma Jinfu and his team started from this point and decided the theme to be "Seven Colors". Sculptures, facilities and decorations are applied to highlight the characteristics of three halls: Absolute Design (Hall 1), Creative Materials (Hall 2) and Deco Trends (Hall 4). Details are carefully designed to create a colorful and diversified exhibition for design.

全国14城市上调首套房利率

11月13日,建设银行北京分行将首套房贷款利率上调到基准利率的1.05倍,申请贷款人如果希望尽快放款,就需要主动上浮利率到基准利率的1.1倍。全国14个城市的部分银行首套房贷款利率不仅告别"打折",还在基准利率基础上上浮5%~30%不等,部分城市甚至有上浮50%的银行。

CHINA 14 CITIES INCREASE FIRST-SUIT INTEREST RATE OF MORTGAGE LOANS

In Nov.13rd, China Construction Bank Beijing Branch increased the first-suit interest rate of mortgage loans to 1.5 times the benchmark one-year lending rate. If lenders wants to make a loan as soon as possible, the interest rate of mortgage loans rises to 1.1 times the benchmark one-year lending rate. A few of banks in China 14 cities stop offer discount for first-suit interest rate of mortgage loans ,but also uplift 5%~30% the benchmark one-year lending rate., some of which even ascend to 50%.

CDA·2012中国设计奖(红棉奖)提名奖出炉

近日CDA中国设计奖(红棉奖)发布2012年度评选提名名单,世界各地856件参赛作品仅有62件获得提名资格,评审之严令业内外吃惊。部分好设计因参评门槛未获提名。
CDA·2012中国设计奖由原"红棉奖"转型而来,是全球第一个以"再定义中国设计"为主题的奖项。CDA·2012中国设计奖(红棉奖)分人居、营商、人物三大类别,主张和重视具有"中国DNA"的优秀设计,要求设计师和设计机构能从研发过程以及产品自身描述自己对中国设计的理解和定位,其参赛作品必须是"融入中国文化或设计者能通过产品阐述中国文化对其设计意义的好设计"。
据悉,作为2012年广州国际设计周的重要内容之一,CDA·2012终评将于12月6日在广州保利世贸博览馆一号馆进行,5位国内评委将会同丹麦设计博物馆公谊会主席约翰·亚当·林伯、IFI主席沙仕·卡安等五位国际评委,现场评选出年度获奖作品,所有提名奖还将同时通过为期三天的公开展示及大众投票选出"公众大奖"。

CDA·THE NOMINEES FOR INNOVATIVE DESIGN AWARD WAS ANNOUNCED

Recently, the nominees for CDA Innovative Design Award was announced, only 62 among 856 entries gained the nomination. The strictness of standards surprised the industry participants . A few of good designs was not nominated because of requirements of participating in judgment.
CDA Innovative Design Award is the first one be themed of " Refining China Design" all over the world. This award is divided into 3 categories– habitations, business, and characters, standing and focusing on excellent designs with Chinese DNA. It requires the designers and the design institutions to understand and position Chinese design from development process and products itself, and their entries submitted must be " the good design that blend in Chinese culture, or designers could interpret Chinese culture and its design meaning through products".
It is said that, as one of important contents of Guangzhou International Design Week 2012 , CDA.2012 final award will be held in Guangzhou Poly World Trade Expo Center, Hall No.1.. There will be 5 domestic judges and 5 international judges including the chairman of society of friends of Denmark Design Museum and the chairman of IFI etc. They will select out the awarding entries, and after all nominated entries are displayed for three days, one of them will award the "Public Award" through a poll of the public.

INFORMATION | 资讯/设计

MP别墅

MP别墅是一座美丽的滨海建筑,占地面呈一个狭长的梯形,房间主体朝南,走道等区域朝北。一层是客厅、餐厅、厨房等公共区域,室内外通过滑动玻璃门分隔,门外是一块露天休闲区,地面上铺了木地板,上方是别墅二层的悬臂结构,提供了遮阳体。地下还有杂物间,车库和综合区。二层分为两部分,分别是儿童卧室和主卧。

MP House
The house, situated on a narrow trapezoid plot of land, achieves the best possible orientation: the main rooms face directly South while the passing areas face North. With the help of light, the project consists of a ground floor that holds public spaces like the living room, the dining room and the kitchen. These are connected with the outside thanks to large sliding glass doors and a wooden paving that serves as a path running towards the porch, which is protected by the cantilevers of the upper floor. The basement contains a utility room, a garage and a polyvalent space. The first floor is divided by a double-height central piece into two different areas: both children's bedrooms on one side and the main bedroom on the other.

巴厘岛竹屋

巴厘岛当地设计师Ibuku,完成了几个大大小小的竹屋,与当地景观和谐的融为一体。这些竹屋由当地的工匠用非常传统的方式建造而成,具有高质量的结构和很长的寿命。

Green Village
Local Balinese practice Ibuku has designed several houses that co-exist with the natural landscape, using bamboo as its only building material on a multitude of scales. The houses are amongst the first structures that use very traditional building methods and local craftsmen that result in a high quality structure and life.

克里科家庭中心

加拿大阿西尼波河公园高耸的榆树与温柔起伏的地面融入了附近居民的集体记忆中,克里科家庭中心暂时植入到这片土地上,并不断调节着自己以适应周围自然的节奏。随着时间流逝,材料的老化与环境的风化作用会使建筑成熟、融入并最终脱离这个环境。

Qualico Family Centre
The towering elm trees and gentle meadows of Winnipeg's Assiniboine Park have for generations formed an inspiring backdrop to the city's collective memory. The Qualico Family Centre connects to the temporality of this landscape, evolving, growing and decaying in harmony with the natural rhythms of its surroundings. As time passes, materials that age and weather allow the architecture to mature, growing into and out of its site.

坎贝尔住宅

俄罗斯亿万富翁Vladislav Doronin委托扎哈•哈迪德在莫斯科为他想象中的超模女友娜奥米•坎贝尔设计一座住宅。这座宇宙飞船一样的住所位于俄罗斯Barvikha,周围被高达20m的松树与桦树环抱,形成一个比较私密的环境。

Spaceship House for Naomi Campbell
Russian billionaire Vladislav Doronin has commissioned Zaha Hadid to design a house in Moscow, Russia for his supposed supermodel girlfriend Naomi Campbell. The spaceship-like abode is located in Barvikha, surrounded by pine and birch trees reaching up to 20 meters high providing a bit of privacy.

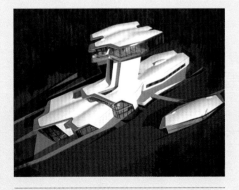

Abbots Way别墅

这栋别墅有五间卧室,周围被成熟的树木和一个小型湖泊环绕,营造出非常放松隐蔽的居住环境。别墅造型非常现代,与英国南海岸美丽的乡村环境相得益彰。

Abbots Way
Abbots Way, the latest creation by AR Design Studio, is a stunning five bedroom house. Bordered by mature trees and a small lake, this spectacular house creates feelings of ultimate relaxation and privacy, whilst its contemporary design juxtaposes superbly with its beautifully rural location, on the south coast of England.

U型住宅

这座山顶住宅是葡萄牙设计师Jorge Graca Costa为一个专业冲浪运动员及其家人设计的。这座U型住宅灵感来源于传统的地中海式院落,以抵御于葡萄牙西部常有的西风。坐落于被树木环抱的山顶,住宅将圣洛伦索湾的美景框入室内,并通过院落调节着当地的地中海式的气候。

U-House
This hill-top house by Portuguese architect Jorge Graca Costa was designed for a professional surfer and his family.The form of the U-House took inspiration from traditional Mediterranean courtyard houses to protect it from the windy climate of western Portugal. Perched at the top of a hill and surrounded by trees, the shape of the house frames the view over the San Lorenzo bay and helps to moderate the climate within the courtyard.

空中庭院

空中庭院是一个约1 858 m²（20 000平方英尺）的会员制俱乐部，功能包括度假别墅、办公和一些娱乐设施。利用院落和坡屋顶的形式来组织功能，在有限的界限内嵌入了好几个庭院，并利用坡屋顶来调整院落间建筑的功能空间，从上方能看到所有的院落。

Sky Courts

Sky Courts is a 1,858 m² (20,000 sqf) corporate club house that incorporate short-term housing, office space, and entertainment facilities utilizing the logics of the courtyard and sloped roof. The project packs several courtyards into a defined perimeter and utilizes the sloped roof to accommodate program in the wedge between courtyards, allowing the project to read as 100% courtyard from above.

济州岛多姆总部

这座建筑是韩国像素房屋工作室为济州岛国际IT公司，韩国著名门户网站多姆公司设计的总部。公司位于远离大都市的岛上，来躲避这个城市化最严重的国家不断增长的人口。设计要求在这个乌托邦一样的岛屿上建立一个创造性工作的社区。

Daum Headquarters

Korean firm mass studies has designed the 'Daum Headquarters' for an international IT company based in the Jeju province, an island off the southern coast of Korea. The company positioned themselves on an autonomous island away from metropolitan areas, to oppose their competitors with an awareness of Korea's population growth and densities representing it as the most urbanized country in the world. The proposal creates a creative work community within a Utopian setting.

秘鲁绿墙

布宜诺斯艾利斯的建筑设计事务所Gonzalez Moix Arquitectura近来在秘鲁首都利马郊外完成了一座多功能的商业和办公建筑。设计将多种材料组合在一起，比如在混凝土中嵌入木材或金属材质。不过这个项目最有意思的地方是三维绿墙，设计师将回收利用的木板随意组装在墙上，并让各种植物从墙的裂缝中萌芽。

Green Wall

Buenos Aires-based practice Gonzalez Moix Arquitectura has recently completed a mixed-use commercial and office building in la Molina, a suburb of Lima, Peru. boasting a raw material palette with a folding concrete structure imprinted with the wooden form work, steel, the project's true treasure is the three-dimensional green wall designed by Veronica Crousse as a mural of recycled wooden planks fitted together like a haphazard puzzle with several kinds of plant species sprouting from the cracks and seams.

CET艺术中心

CET艺术中心位于多瑙河岸边，采用的曲线形状像是一条鲸鱼，为左右两边的建筑建立了视觉上的联系，河边的平台为当地居民提供了休憩的场所，像是这座建筑对多瑙河的回馈。

CET Art Center

Standing for 'Central European Time' and the Hungarian word for whale, this structure on the banks of the Danube takes on the curving form of a whale's body. Establishing a visual connection between the physically separated sides of Buda and Pest, the site returns riverside terraces to the city for the residents to frequent.

圣那泽尔剧院

法国圣那泽尔剧院紧邻着第二次世界大战中被毁坏的新古典主义火车站。这座巨大的实体建筑形式源于附近的碉堡，混凝土的表面上印上了镂刻的花纹图案，图案花纹来自17世纪的丝质纺织品。

Théâtre of Saint-nazaire

The 'Théâtre of Saint-nazaire' in Saint-nazaire, France, adjacent to remnants of a neoclassical train station destroyed in World War II. The monolithic forms takes cues from a nearby bunker and the concrete surface is stamped with a perforated floral pattern derived from motifs of 17th century silk textiles.

UBC药学系馆

医药科学及研究开发中心位于英属哥伦比亚大学(UBC)的心脏位置并位于学校重要的入口处，考虑到这点，建筑的形态类似校园东南大门，地面层开放、透明、具有亲和力，对外展示着建筑的公共功能。

Faculty of Pharmaceutical Sciences and Centre for Drug Research and Development (UBC)

The Faculty of Pharmaceutical Sciences and Centre for Drug Research and Development is sited in the Heart of UBC's campus.and is located at an important university entry point. With this in mind the building has been designed to act as a gateway to the southeast edge of the campus, engaging the community with a ground floor that will be open, transparent, inviting, and one that will showcase the public function.

ELEGANT BRITISH-STYLE ARCHITECTURE AND ROMANTIC GARDEN LANDSCAPE | British Villa, Chengdu

典雅高贵的英伦建筑 风情浪漫的园林景观—— 成都中德英伦联邦

项目地点：中国四川省成都市	Location: Chengdu, Sichuan, China
开 发 商：四川中德世纪置业有限公司	Developer: Sichuan Zhongde Century Properties Co., Ltd.
建筑设计：广州宝贤华瀚建筑设计事务所	Architectural Design: Baoxian Huahan Architects, Guangzhou
景观设计：泛亚国际	Landscape Design: EADG
占地面积：600 000 m²	Land Area: 600,000 m²
建筑面积：2 200 000 m²	Floor Area: 2,200,000 m²
容 积 率：3.6	Plot Ratio: 3.6
绿 地 率：55%	Greening Ratio: 55%

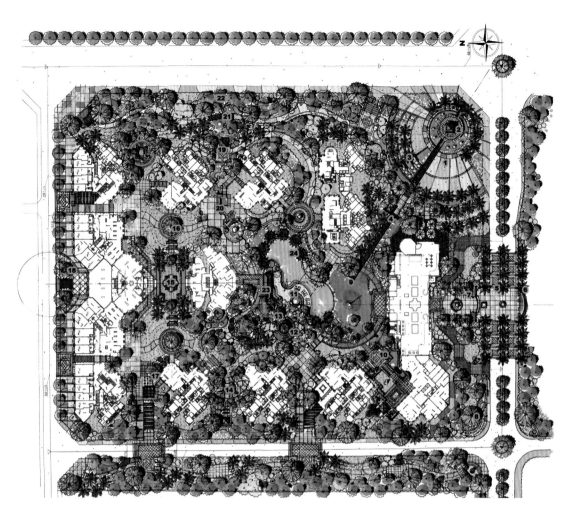

总平面图 Site Plan

B区园林总平面图 Landscape Plan for Zone B

MASTER AND MASTERPIECE | 名家名盘

项目概况

中德英伦联邦位于成都国际城南天府大道南段，地处高新领馆区核心区域，北临5平方千米的西部金融中心——金融城，东接规划为10平方千米的花园式新川创新科技园区。天府大道、红星路南延线、地铁1号线、6号线均从项目经过。作为入驻成都的开篇力作，项目秉承"关爱有家"的人本思想，整合国内外精英团队和优良供应商，以颠覆传统豪宅的价值标准，通过精益求精的奢华品质、风情浪漫的园林景观、典雅高贵的英伦建筑，致力于实现青年才俊的高品质生活梦想。

规划布局

项目总规划用地60多万平方米，总建筑面积约200多万平方米，致力于打造国际城南首屈一指的超级大盘。小区布局合理，人车分流，彰显人文关怀；三重豪华入户大堂，尊贵私家奢享。为小区业主量身度造的约6 000 m² 高级商务会所、8 000 m² 主题会所，集景观游泳池、恒温游泳池、英式下午茶、休闲运动中心、咖啡屋于

一体。独具匠心的首层局部架空,与英式花架连廊迂回贯通,打造专属青年才俊的"梦想泛会所",堪称天府新城高尚英伦风情主题社区典范。

建筑设计

项目采用围合式穿插点式布局,为实现较高的绿化率,降低建筑密度,局部采用超高层建筑,户型朝向以最大景观化为原则,有效解决高层对视问题。还特别设计首层下沉式私家花园,270度八角豪华观景卧室,顶层梦想创意阁楼,与小区数十万平米英伦风情园林交相辉映。项目以英伦文化与生活气韵交融的古典风情为蓝本,精心打造为涵盖住宅、高级会所、主题商业、双语幼稚园等多种业态的高端国际社区。

景观设计

结合项目建筑英伦风格的特色,项目的景观设计突出以英伦文化与当地生活气息相结合,着力打造极具英伦特色的浪漫风情园林。高达55%的住区景观绿化率,也为整个社区良好的生活环境提供了保障。

户型设计

本项目户型设计理性,注重实用性、舒适性及私密性。主力户型区间为60—120 m^2,除70—90 m^2梦想幻变主力户型外,更有60 m^2左右精致舒居户型,90—120 m^2,豪华观景户型,户均赠送率超过40%。

MASTER AND MASTERPIECE | 名家名盘

S 户型图 Layout of S Type

G 户型图 Layout of G Type

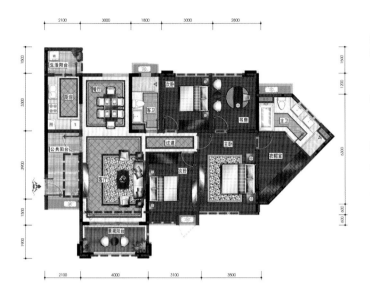

K 户型图 Layout of K Type

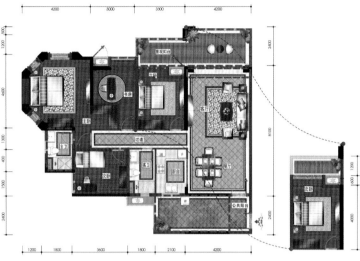

C 户型图 Layout of C Type

MASTER AND MASTERPIECE | 名家名盘

Profile
Located on the south section of Tianfu Avenue of Chengdu City, within the high-tech parks and consulate area, the project British Villa faces 5,000,000 m² Financial City in the north and 10,000,000 m² Xinchuan Sci-tech Park in the east. Tianfu Avenue, south extension of Hongxing Road, Metro Line 1 and Line 6 are passing through the site. As the first development by Zhongde in Chengdu, British Villa adheres to the idea of "caring family", commissions to the world-famous architects team and adopts high-end building materials, trying to provide a luxurious, romantic and elegant British-style neighborhood for quality life.

Planning and Layout
With a total land area of 600,000 m², the project features a total floor area of 2,000,000 m², envisioned to be the top residences in south Chengdu. With clear layout, walkers and cars are separated to ensure safety; luxurious entrance hall provides supreme honor and privacy. The customized 6,000 m² high-class business club and 8,000 m² theme club consist of landscape swimming pool, thermostatic swimming pool, British-style tea house, fitness and sports center, coffee house, etc. The unique elevated ground floor connects with the British-style flower corridor, providing a dream club for the clients.

Architectural Design
Buildings are arranged in enclosed layout, with some alternately in between. To get higher greening ratio and low building density, some buildings are designed with super heights to get maximum landscape views. There are

also sunken home garden, 270-degree landscape bedroom, and dream loft complementing the large area British-style gardens. With British cultural and living atmosphere, it tries to build a high-end international neighborhood integrating residences, high-class clubs, theme commerce, bilingual kindergarten, etc.

Landscape Design

The landscape design is designed according to the architectural style, combining British culture with local culture to create exotic romantic garden landscapes. The greening ratio is high to 55%, providing an ideal living environment.

House Layout Design

The design of the house layout pays attention to functions, comfortableness and privacy. There are many types of unites with a floor area ranging from 60~120 m^2. 70~90 m^2 units are changeable for different requirements, 60 m^2 small units are comfortable, and 90~120 m^2 units are luxurious to get great landscape views.

建筑的美学价值、时代特征与可持续发展
——访北京奥思得建筑设计有限公司董事总经理 杨承冈

■ 人物简介

杨承冈
北京奥思得建筑设计有限公司董事总经理
国家一级注册建筑师
法国布列塔尼建筑学院硕士
清华大学建筑学院，建筑学学士

杨承冈先生毕业于清华大学建筑学院，一直致力于融合东西方设计理念和社区哲学研究，营建具有中国特色的新时代建筑艺术。在高级居住建筑、商业综合体、度假区、办公建筑等领域，结合绿色生态技术的多方面应用，努力创造出富有空间变化流动性，传承城市建筑文脉，将自然环境和建筑完美结合的多种类型的区域建筑，风格富于创新，收到广泛好评，在建筑应用和社会美学方面能做到很好的结合。

■ 代表项目

先后主持北京中海雅园高尚社区、首都师范大学外语教学楼、北京亮马名居高级公寓、北京乐府江南居住区、北京万柳文化公园、北京香江雅园别墅区、郑州中原杏湾城时尚社区、河北西柏坡干部学院、阿尔及利亚国家综合演出厅、北京芸溪境分时度假村、呼和浩特东岸国际B区、呼伦贝尔珊瑚墅别墅区、山西尧庙国际商贸城、宁夏固原尚都国际商贸城、葫芦岛北港工业区管委会办公区、鄂尔多斯康巴什左岸公园别墅区等多个项目。

《新楼盘》：建筑往往具有时代感和历史性，您是如何看待的？在设计过程中怎样去把握这种时代感？

杨承冈：建筑是社会历史发展的承载物，它的表现是综合性的，不同的发展历史时间段，人们评价建筑的空间、装饰、环境，会随着不同的感受、认知、体验发生变化，但建筑的核心理念和价值是始终延续的。传统的中国建筑由于受到中庸的儒学文化，和相对内敛的慢生活方式影响，从大型宫廷建筑到普通民居均比较封闭，相互之间的交流不十分充分，但重视人和自然的关系。现代社会人与人的交往大幅度增加，生活节奏加快，空间的流动性要求提高，城市空间变得越来越开放，建筑的多样性不断增加，原有的城市格局被打破，自然环境被压缩，城市公共空间、人居、自然景观被割裂开，放置在不同的区域，而不具有原来的共生感，这就是现代建筑造成的后果，也就是所说的建筑时代特征。

我们所处的城市也被快速生长的各色建筑冲击着，被城市核心区超过两、三百米的超高层建筑无形压迫着，也被许多怪异的大体型的建筑物震动着，但建筑设计师有时更应当跳出这样的节奏去思考，不能完全被这样的时代发展牵着鼻子走。现代城市不应盲目发展这样的建筑，我们应该把握住建筑应有的原则，设计时应该做到"适合一个城市的文化环境，适合一个城市发展的趋势，适合建筑功能的需要，同时适合建筑的特征"，在此基础上再试图做到适度的超越，既不过多地求新求怪，但要能够充分展开空间的想象。我们的设计注重建筑空间的组合和流动，注重城市文脉的传承和持续发展，把人文的因素放在首要位置，我想这是最符合时代要求的。

《新楼盘》：谈一下国内外项目在设计上有何不同之处？

杨承冈：无论是国内我们接触的各种类型的项目，还是在北非曾经面对的那些荒漠上营建的新城、阿拉伯地区拥挤城市空间状况下的改造项目，或者是在加拿大为冬奥会服务的运动员村，我们的设计团队实际上一直秉承着相同的设计理念，即对建筑的理解，对城市空间和社会形态的相互关系的理解。建筑不是设计手法的堆砌，而应当是空间的流动变幻传递出的不同的意识形态。如果要分析出国内外项目的不同，应该是我们的设计中十分重视的社会文化性和独特意识性。不同的国家有着他们各自发展的轨迹，独具特色的文化表现性和人群相互间交流的方式，这其中最有代表性的就是语言。我们正试图象语言传递一样，用建筑文化的语汇来表现出每一个项目所在地域的历史文化传承，我们希望设计能够给出很好的城市延续发展的保证，而不是粗暴的割裂。就像在山西临汾设计的锦悦城国际商贸城，和阿尔及尔的国家综合演出厅项目，我们重点研究的不仅是建筑本身的功能，更重要的是场地环境，城市文化的特征，和地区居民交流的方式，我们要把这些吸收融合，重新发酵表现在作品里，这样才能保证这个项目适合它所在的城市，适合这里的人。某种意义上建筑师应当更多地把视野放在城市历史发展的角度去思考。

《新楼盘》：你们的业务涵盖了包括建筑、城市、景观等多个方面，谈一下建筑设计与景观设计之间的关系以及区别？

杨承冈：城市、建筑和环境，三者应该是共生的，建筑不能脱离它所处的环境，也不可能超越它的城市生长。作为一名建筑设计师，应当做到"把握好城市发展的脉搏，把握好场域环境的特征延续，把握好建筑功能和空间的合理性"，这三点做好了，建筑的生命力就自然而然地具备了。因此，在我们公司的设计工作中，十分注重全方位地把控这三个要素，我们不希望只是孤立地进行其中一个环节，而是通过规划、建筑、景观团队，来达到对一个项目的完整设计。因此很难说这三个环节：规划设计、建筑设计和景观设计具有多么大的本质不同，它们是一个设计作品的三个不同表现层面，而核心理念应该是共同的。

《新楼盘》：介绍1~2个你们最近的设计项目，谈谈它们各自的特点和风格。

杨承冈：我们公司长期以来一直专注于高端的风格居住建筑的研究，如我们在呼和浩特市做的东岸国际社区的核心就是重点营造高雅的居住文化理念和强调自然生态的居住环境。通过对道路交通体系的全新构建、绿化景观的有机梳理，使之成为一个完整和谐的统一体，再在此基础上规划各个功能组团

将不同类型的居住单元和建筑空间合理地串接起来，做到真正意义上的情景交融，让建筑自然地追随景观环境。设计中着力于造景与交通相结合，意图在低密度居住社区中合理经济地建立一种人车分流的全新交通体系，从而使居住者能够真正拥有宁静安详的高尚社区生活，通过社区东、南、北方向出入口，机动车直接进入半开敞地下车库，极小地缩减了机动车进入社区地面的路段，保证区域内地面上整体道路中无机动车通行，使得不论是高层公寓、还是多层洋房和别墅的停车和出行全部在半自然地坪下解决。规划注重建立"生态山水之城"的概念，强调建筑和环境的融合，在居住建筑的规划上运用新城市自然主义的设计方法，力求使建筑如自然生长在景观空间周围，结合地形起伏和水体的走向，曲折布置，控制对景观带的开放程度和朝向角度，做到相互交融促进。

公司另一重要设计研究是城市公共建筑，包括大型商业综合体和文化建筑，在这个领域，我们充分发挥城市研究的特长，从城市宏观发展的层面去定位建筑的特质，切实做到建筑在城市中的有机生长。我们近期在山西临汾设计的奥体中心，在充分考虑临汾自唐尧以来悠久的文化历史，充满活力和前景的城市发展趋势，以及体育运动富于动感特质的基础上，确立了建筑的定位和主要理念。整体造型充分考虑采用了充满流动感的曲线形态，汲取临汾地区龙山文化的代表——陶器和玉器的造型和色彩，在建筑细节上运用了现代主义的手法，结合涂料和铝合金装饰，大面积的落地玻璃条窗，简洁而富有朝气，营造出未来主义色彩的建筑风格，富于进取心和时尚感，形成层次丰富、错落有致、气魄宏大的现代建筑群。周边的汾河景观和主体建筑做到最佳结合，方案精心营造了"一水、一圆和五环"，将建筑组群设计出跃动的美感。

《新楼盘》：结合当下建筑行业的发展，谈一下未来生态建筑是怎样的一个发展趋势？

杨承冈：生态建筑当下谈论的越来越多，也越来越受到各方面的重视，这的确是建筑未来发展的一种趋势。一个建筑是否生态，一座城市或一个社区是否生态环保，它包含硬件和软件两个层面，硬件主要体现在建筑材料和构造的改良，建筑节能技术，相对而言是比较容易达到和显现出来的；软件层面更多地是体现在人们的意识形态当中，这包括社区人群的生活习惯和方式，建筑和环境的再生利用，城市模式的可持续生长发展等等，相对硬件来说要困难许多，它必须经过长时间的培养、营造才能慢慢得以形成。人们可能很容易拥有良好的物质条件和节能设施，但不一定养成最佳的生活方式，这需要我们在建筑规划设计中加以规范和引导。

我们近些年来通过与境外设计师的协作，对生态的理念有了一些体会，并不断地努力应用在我们的设计作品中。我们在北京郊区和河北等地风景区尝试做过一些度假酒店和别墅建筑，完全依托于原始地貌，不破坏当地自然形态，采用本地的石材，不过多采用不可再生的混凝土材料，对污水进行生化处理，雨水再生利用，在软硬件两个层面均做到了生态的持续营建。

一座建筑的建成，一个区域场所的形成，势必会约束到人群的行为方式，这就需要每一个建筑设计师自身要具备良好的生态意识。我们设计的建筑是应当不破坏生态环境，不与城市文化模式形成重大冲突的，它是将一个城市的文脉赋予了一定意义上的时代特征，这是建筑持续发展的生态要求。

《新楼盘》：请您谈一下建筑的可持续发展对于城市发展的意义。应该如何通过低碳生态的设计理念去实现城市的可持续发展？

杨承冈：很高兴关注到建筑的可持续发展这一主题，我在法国留学期间研究的就是建筑的可持续发展和利用。相对而言欧洲的城市普遍十分重视城市文化的延续发展，注重城市的记忆，而国内由于出现过若干次建筑思想的断层，又经历了过多的超正常速度的膨胀式建设，因而众多城市的肌理和文脉遭到了破坏，十分可惜。现在我们城市的模式几乎雷同，如果短时间内经历几个城市，人们会不自觉地产生恍惚感。大片的旧城区由于人口密集，房屋众多，很难安置而被搁置，与此同时，政府和开发商都更愿意去规划和开发城市的新区，这造成了大量的盲目建设和不必要的浪费，城市本身的特征和形体也遭到了极大地破坏，很多的破坏是完全不可能补救的，非常令人痛心。其实低碳生态的理念更应当运用在城市合理发展的过程中，许多旧城区经过合理地规划是完全可以焕发出生机，产生新的价值。北京的胡同四合院不是被拆除，而是更新利用，赋予新的功能；一些废弃的厂房建筑如果被相对集中地保留治理，是完全可以具有新的价值；湖水及其周边的各种形态的植被应该尽可能地被循环保护，而不是被建设，同时这些具有历史记忆的建筑可以很好地起到文化传承的作用。人们总是愿意从回忆中产生时光的穿梭，作为建筑设计师应该帮助人们对城市，对这里的环境和曾经发生的事情，留有完善的记忆，这样才能做到最佳的持续发展。

房屋建造是一种工程美学，它集合了众多技术，包括生态环保的种种措施，但是成为真正的属于这座城市的建筑，需要倾注时间的沉淀，良好的空间对话和对人文态度的反映。我们一直希望我们的作品是能够成为一个城市的自然肌理和文化延续的。

阿尔及利亚演出厅

东岸国际

乐府江南

The Aesthetic Value, Characteristic of Epoch and Sustainable Development of Architecture

—— Yang Chenggang, managing director of BHAD

Profile:

Managing director of BHAD
National first-class registered architect
Master of Architecture of Ecole D'architecture de Bretagne (The Brittany National College of Architecture)
Bachelor of Architecture of Tsinghua University

Mr. Yang Chenggang graduated from the School of Architecture, Tsinghua University, has been committed to the integration of Eastern and Western design philosophy and community philosophical studies, constructing architectural art of new era with Chinese characteristics. He applied green eco-technology in various projects like high-end residential building, commercial complex, holiday resort and office building, which are full of space variation, inheriting urban context and combining natural environment with architecture perfectly. And he also received the wide acclaim in combining architecture with social aesthetics.

Representative works:

Beijing COE Haiya Garden Community, Foreign Language Teaching Building of Capital Normal University, Beijing Landmark Palace, Beijing Yuefu Jiangnan Residential Area, Beijing Wanliu Cultural Park, Beijing Xiangjiang Graceland, Zhengzhou Zhongyuan Apricot Bay City, Hebei Xibaipo Cadre College, Algeria National Comprehensive Performance Hall, Beijing Yunxijing Fenshi Holiday Resort, Hohhot East Coast International Area B, Hulun Buir Coral Villa, Shanxi Yao Temple International Business City, Ningxia Guyuan Shangdu International Business City, Hu Lu Island North Port Industrial Area Management Committee Office, Erdos Kangbashi Left Bank Park Villa, etc.

New House: Building always has a sense of contemporaneity and history. What do you think of it? How to grasp the sense in design process?

Yang: As the bearer of social and historical development, the performance of building is integrated, in different historical period, people evaluate architectural space, decoration and environment with different feelings, cognition and experience, but the core concept and the value is always continued. Due to the doctrine of Confucian culture and relatively restrained slow lifestyle, traditional Chinese buildings from palace buildings to ordinary houses are relatively closed, exchanges between each other is not very full, but they emphasis on the relationship between man and nature. Social interaction between modern people is increasing significantly, further requirements are made on space liquidity, and urban space is becoming more and more open, architectural diversity is increasing, the original urban pattern is broken, the natural environment is compressed, urban public space, habitat and the natural landscape are separated, placed in a different area, rather than with the original sense of the symbiotic, this is the consequences of modern architecture, which is said to be the architectural characteristics of the times.

The city we lived in is immersed with the rapid growth of all kinds of buildings, such as ultra-high-rise buildings more than two or three hundred meters in the urban core area and the weird giant volumes, but architects should be able to think out of the rhythm sometimes and can't be completely led by the development of this era. Modern city should not evolve without consideration like that. We should grasp the due principles, what we do should be appropriated for the cultural environment of the city, the development tendency of the city, the requirement of building function and the building features. And we can try to exceed on that basis, not pursue novelty and weirdness too much, but to be able to fully expand the imagination of space. Our design focuses on the combination and circulating of architectural space, attaches importance urban context heritage and sustainable development and puts the human factors in the first place. I think this is the most in line with the requirements of the times.

New House: What are the differences between your designs at home and abroad? Can you talk about that?

Yang: No matter the various domestic project we are in contact with, or facing the construction of new city on the North Africa desert, or the reconstruction project in the crowded urban space in Arab region or the athletes village for Canada Winter Olympic Games, our design team has actually been adhering to the same design philosophy, i.e., the understanding of architecture and the understanding of mutual relations between urban space and social pattern. Building is not the concoction of design techniques, but the different ideologies that conveyed by the changeful circulating of space. To analyze the difference between the projects at home and abroad, our design attaches great importance to the socio-cultural and unique consciousness. Different countries have their respective development path, unique cultural expressions and the way to communicate with each other, and language is the most representative. We are trying to pass like language, to show the historical and cultural heritage of the geographical location of each project with the vocabulary of architectural culture, we hope to be able to ensure the continuation of urban development well, rather than the brutal fragmented. Take the international business city in Linfen, Shanxi and Algeria National Comprehensive Performance Hall as examples, we focus on not only the function of the buildings themselves, but more importantly the site environment, cultural characteristics of the city and the way we communicate with local residents, we have to absorb these and present them in our works, so as to ensure that the project is suitable for the city and the people. So, in a sense, architects should pay more attention to the historic development of a city.

New House: Your business covers many aspects such as architecture, urban planning and landscape. What are the relation and differences between architectural design and landscape design?

Yang: City, architecture and the environment, all three should be symbiotic. The building can't be separated from its environment, and can't go beyond its urban growth. "Grasp the pulse of urban development, grasp the characterized continuation of the field environment, grasp the rationality of architectural features and space", as an architect, you should do well in these three, then the vitality of the building comes along naturally. Therefore, in our design work, great and comprehensive attention is to give to all three elements. We do not want any part is isolated, but to complete a project design through planning, architecture and landscape teams. It is difficult to say how planning design, architectural design and landscape design are essentially different, they are three different performance levels of a design works and they share a common core concept.

New House: How about introducing one or two of your recent design projects? Talk about their features and styles?

Yang: Our company has long been focused on high-end style residential buildings, as the international community we do in Hohhot, is focusing to create the elegant living culture concept and emphasize the natural ecological living environment. New road traffic system and organic green landscape combine into a harmonious unity. Based on various functional groups, the different types of dwelling units and building space concatenate properly to follow the natural landscape environment. Planning focuses on the establishment of the concept of "ecological landscape of the city," emphasizes on the integration of building and environment, uses new urban naturalism design method, and strive to make the building just like something that grows around the landscape naturally. It combines with the terrain and the trend of water body to control the openness and orientation of the landscape areas, promoting each other mutually.

Another important design research is the city's public buildings, including the large commercial complexes and cultural buildings, in this field, we give full play to the strengths of urban studies, to find out the characteristics of the buildings from the macro-level development of the city and to make sure that the buildings in the city are in organic growth. Recently, we have designed the Olympic Sports Center in Linfen, Shanxi, on the basis of fully understanding its long cultural history, vibrant prospect and rich dynamic characteristics, we established the positioning of the building and main idea. Overall shape gave full consideration to the using of curve lines with a sense of movement, learned from the representatives of Longshan culture—the shape and color of the pottery and jade. In the architectural details, modernist approach is used, a combination of paint and aluminum alloy decoration, a large area floor-to-ceiling glass bar window, concise and full of vigor, creating a futuristic architectural style, full of entrepreneurial spirit and sense of fashion. In addition, the Fenhe River landscape around and main building achieved the best combination that expresses the beauty of the design.

New House: Combined with the current development of the construction industry, what do you think of the future trend of eco-building?

Yang: Eco-building draws more and more attention at the moment, and it is indeed a trend for future building development. Whether an ecological building, or an eco-friendly city and community, they matter to two levels, the hard one and the soft one. The former one is mainly reflected in the improvement of building materials and construction, building energy-saving technology that is relatively easier to reach; the soft level is reflected in the ideology of the people, including living habits and ways of community populations, recycling of building and environment, sustainable growth and development of urban model, it is much more difficult to realize and it takes time to foster, to create and to shape. People might easily to a good physical condition and energy-saving facilities, but not necessarily to develop the best way of life, we need to regulate and direct in a certain sense in architectural planning and design.

Collaboration with foreign designers in recent years helps us to better understand the ecological concept, and we constantly strive to apply the concept to our design works. We used to design some resorts and hotels in Beijing outskirt and the scenic area in Hebei, we completely rely on the original landscape, not to destroy the local natural form, use local stone and a good amount of nonrenewable concrete materials, dealing the wastewater with biochemical treatment and recycling the rainwater, which realized the continuous construction of ecology in two levels.

The completion of a building, the formation of a regional place, is bound to restrict the behavior of the crowd, which requires every architect to have a good ecological awareness. Our design should not harm the environment and not conflict with the city cultural pattern. It endows the urban context the characteristics of the times in a certain sense, which is the ecological requirements of sustainable development of buildings.

New House: Could you please talk about the significance of sustainable development of the building to a city? How should we to achieve the urban sustainable development through low carbon and ecological design concept?

Yang: I am pleased to be concerned about the sustainable development of building, that's what I have researched in France. European cities generally relatively attach great importance to the continuation of the development of urban culture and focus on the city's memory. However, it is a pity that urban contexts of most domestic cities have been destroyed by the several architectural discontinuities and supernormal expansion. Now, most of the city models are essentially the same. Large tracts of the Old Town were shelved due to the dense population and masses of houses, at the same time, the government and developers are more willing to plan and develop the new district of the city, which caused a lot of blind construction and unnecessary waste, and the urban characteristics and the city itself have also been greatly damaged, a lot of damages can't be remedied, very distressing. In fact, the concept of low-carbon and eco should be used in the process of rational urban development. Many Old Towns can spring to life after a reasonable planning and then produce new value. Beijing's Hutong courtyards were not removed, but updated and being given new functions; if the abandoned factory buildings were improved and reserved intensively, it is entirely possible to have a new value; various forms of the lake and its surrounding vegetation should be protected as much as possible rather than being constructed, at the same time, these buildings with historical memory can play the role of cultural heritage well. People are always willing to shuttle time from the memories, so architects should help them to keep the memory of the environment and what happened to the city, thus to achieve the best sustainable development.

Housing construction is an engineering aesthetics, it is a collection of a number of technologies, including various measures of eco-environmental protection, but to become the building that truly belong to the city, it takes time, requires favorable dialogue between spaces and reflection on humanistic attitude. We always hope that our work is able to become a city's natural texture and cultural continuity.

营造是设计与施工之间的有效推进剂

——访广州市圆美环境艺术设计有限公司技术总监 吴应忠

■ 人物简介

吴应忠
广州市圆美环境艺术设计有限公司 技术总监
肇庆科技信息产业园——华南智慧园景观设计总规划师

吴应忠先生针对景观设计图纸到实体景观施工管理的过程，提出"高效实现景观设计方案，营造是设计与施工之间的有效推进剂"理念，主张设计与施工结合，营造最适合的景观。充分考虑甲方经济、营销需要，提前考量方案设计在施工中的难点，熟悉园林材料特性、价格及用法，熟悉园建工艺做法，在钢砼结构、木结构、水系统、植物造型布置方面均有丰富经验，同时在园林艺术的尺度把握上有独到审美，最大限度地呈现设计方案的精神。

《新楼盘》：请简单介绍一下你们目前在肇庆科技信息产业园的基本项目情况？

吴应忠：华南智慧城位于肇庆端州区，占地面积约20.6万平方米，方向是打造成为华南地区集软件、信息、创意、科研、金融、外包等综合服务为一体的生产性服务业首善新城。园区内规划商务办公、总部办公、创意办公、服务外包、技术研发、商务配套、主题商业等多样的产业发展模式。我们的工作主要是对华南智慧城的总体景观进行规划，并按其发展次序进行逐步深化设计，并驻场跟进施工，保证景观营造的高效高质完成。

《新楼盘》：整个产业园应该是有一个总体规划在先，那么作为景观规划，你们是如何介入，并扮演什么角色？

吴应忠：整个产业园的规划是由广州市城市规划勘测设计研究院完成的，作为景观规划，我们肯定是要以尊重建筑规划设计、建筑立面风格为前提的，相对于规划的"广"、建筑的"硬"，园林景观的特点就是细和柔，通过景观的手法，可以更加烘托出建筑的艺术感，软化建筑棱角冷感，活化整个园区的气氛，提升项目的可鉴赏性，并为营销需求锦上添花，这是我们作为景观设计在每一个项目中的工作原则及重点。

《新楼盘》：请您谈谈整个产业园的景观规划特色？

吴应忠：华南智慧城位于山水城市肇庆，首先这个是大环境特色，那么我们的园林风格绝对是它的一个缩影，山水的、生态的、现代中式的是整个设计的三大元素。我们尝试从鼎湖山水、端州墨砚、粤西南文化等各方面进行方案探讨，反复推敲，并最终融汇进景观手法上，通过具体的形式去表现出来。目前已完成的创意办公区，以"哲思园"为主题，在空间划分、单体立面、绿化构景方面，都是以干练、抽象的手法为主，目的是营造出安静中休闲，思索中创意的景观环境，阳光茶座、装饰景墙、创意植物景门、切割植被、特色采光井、自然石头形状座椅等等，分布园内，整体感觉有景看而不喧闹，清新阳光易近人。

华南智慧城里还有一大特色，就是规划了一个面积两万平方米的人工湖，里面有临湖别墅区、湖中岛屿、湖岸单车环道、亲水台阶等等，结合特色滨水植物、乡土特色植物，湖光山色相辉映，好一派湖区绿林生态公园景观风光。目前项目的推进比较理想，我们也很有信心去把这个项目做好，毕竟，这么大的综合性项目是很难得的，作为一个设计师，真的很荣幸能加入，并会努力完善好我们的设计和服务。

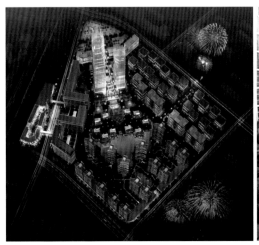

《新楼盘》：营造是设计与施工之间的有效推进剂这句话应该怎么理解？

吴应忠：之所以提出营造，是因为景观设计不仅仅限于设计，用哲学的话说，它是物体，是变化和运动着的，目前在景观项目中，存在着设计与施工脱节的情况，所以应该把营造的重要性提出来。下面我分三点来说：

首先，景观园林的实地施工弹性较建筑大，需要设计师进行工地实践，比较新旧技术水平，因地制宜，与时共进；目前来说，防水技术有很多种，作为设计师，你不知道现场的情况，又怎么会知道用哪种方法最佳呢？

其次，每个项目的施工自然环境、人为环境各异，唯有进行有效指导施工，才能减少返工、节约材料；再有就是根据园林景观的特点，结合设计风格要求，灵活利用特色的搭配达到好的效果，而不是一定采用高档材料。大家都知道，作为发展商来说，始终会计算投入和产出的比例，甚至有些甲方希望以最低的成本去做更多的事。我自己的观点是"好的设计，有效的营造，是为项目降低成本，减少浪费的最佳途径。"

最后是利用设计师的天生艺术感，进行营造，使设计最终更完整地体现在施工成果上，人工建造却与环境相容，富有美感。这点在实际操作上比较多体现在空间过渡、选大型景石、堆山理水、植物组景、小品点缀方面。室内装饰设计现在已划分为硬装和软装两部分，园林景观也应该细分出来，加大对后期营造的重视，方能把景观设计这个专业的重要性提升到一个更高的认可层面。

唯有通过这三方面，甚至还有其他更多的营造方法，我们才能很好地完成一个作品。

《新楼盘》：那么在项目的现场跟进上，你们是怎么去配合甲方？

吴应忠：作为华南智慧城这样一个大型产业园项目，甲方和我们都意识到营造的重要性，所以我们有设计人员专门驻场，并根据项目的进展情况调派各专业设计师去指导，为的就是保证到整个项目的连贯性、整体性，同时施工过程中出现任何问题都能及时了解，并快速处理，这就是设计就在家门口的好处啊。事实上，现场土方、建筑也是在同时进行，二期三期四期也不断推进中，景观营造在这样的项目里不能一步到位的，如同下围棋，既要看得远，又要想得深，才能取得胜利。在华南智慧城项目里，景观设计是我们的主要工作，我们除了对设计精益求精，还需要按照甲方需要，在专业范围内提出合理化建议。比如单就是一个人工湖的湖底驳岸处理，我们就出了多个方案，并请了生态水处理公司、水利防洪专家等来把关节点问题。另外，还要不时参与材料供应商会议，并和集团成本采购部、营销策划部等经常沟通，保证了对项目前后关系的了解，和甲方的各项指令的掌握，从而更好地指导工作。

《新楼盘》：大型产业园的景观设计与一般的住区景观相比，有何不同？要注意哪些因素？

吴应忠：功能决定特色，大型产业园的景观设计在符合整体规划建筑的基础上，由于其使用功能，适应人群，决定了其景观设计是以简洁线条、质朴材质、休闲与生态绿地结合，适可而止的体现产业园特色的小品雕塑为设计主旨。况且智慧园是一个高技术的产业园区，那么，我们要将更多的智慧闪光、思索空间、艺术灵感留给人，我们把园林景观环境做好了，白领精英们在其中能放松心情，汲取大自然的能量，从而焕发更大的工作激情。景观设计其实百变不离其宗，一些基本的准则共通的，只不过针对不同类型的项目，或者不同的项目，注意一个度的把握就可以了。

《新楼盘》：那么大型产业园的景观设计与工业园区景观相比，又有何不同？

吴应忠：我曾经设计过一个工业园区，是台山富华重工工业园，单绿化面积就有20万平方米，园区内有办公楼、厂房、宿舍、泳池会所、贵宾别墅等，但这种工业园区面向对象是本园区的人群为主，主要目的是以绿化为主，建设一个新型工业园为目的，让员工有好的工作环境，让来宾对工业园有好的印象，所以它的开放性相对来说是比较少的。而像智慧园这种大型产业园就需要营造一个富有特色的景观环境，既满足自身需求，同时要吸引更多的企业、群体进来，并依靠各自的特色去整合协调，逐步提升。而这样的景观，开放、时间长、变化大，的确需要设计师的长期跟进，并和甲方管理人员共同用心营造。

Creation is An Effective Propellant between Design and Construction
—— Interview with Wu Yingzhong, Technical Director of Guangzhou Yuanmei Design

Profile:

Technical Director of Guangzhou Yuanmei Design
Chief Landscape Designer of Zhaoqing Sci-Tech Information Industrial Park—South China Wisdom Garden
Wu Yingzhong has proposed that creation is an effective propellant between design and construction to efficiently realize the landscape design from the design drawings to the construction and management. He asserts to integrate the design and the construction for most appropriate landscape. He takes full consideration about the economic and marketing demands of the owner, and the construction difficulty of the plan design. He has well learnt about the characteristics, price and use of garden materials, and the garden construction. He also has rich experience in wooden structure, water system and plant shapes arrangement etc. He has unique ideas about garden art and could maximize the significance of the design plan.

New House: Could you simply introduce your project in Zhaoqing Sci-Tech Information Industrial Park?

Wu: South China Wisdom Garden is located in Duanzhou District, ZHaoqing, with a land area of 206,000 m². It is designed with the goal of a productive service town in South China with a collective service of software, information, originality, scientific research, finance and outsourcing etc. The industrial development modes in the garden is planned to be various as business office, headquarters office, originality office, service outsourcing, technology R&D, commercial supporting and theme business etc. Our work is to present a master landscape plan of South China Wisdom Garden, and further the design step by step based on its development sequence. We will also follow the construction in-site to make sure the landscape is realized in high efficiency and quality.

New House: The whole industrial garden was built under a previous master plan, then how and what do you act in the landscape planning?

Wu: The master plan of the whole industrial garden was finished by GZPI, so we start the landscape planning based on the architecture planning design and facade style. In comparison with the broadness of the master plan, the toughness of the architectures, the garden landscape is featuring softness and gentleness. It could help express the art sense of the architectures, soften the coldness of the architecture edges and corners, activate the atmosphere in the garden, increase the appreciativeness of the project, and contribute to the marketing, which are the working principles and essential points in every landscape design project.

New House: Could you share with us the landscape planning characteristic of the whole industrial garden?

Wu: South China Wisdom Garden is located in Zhaoqing, a city with mountains and waters, which is a general environment characteristic. As an epitome, the garden design is filled with three main elements: landscape, ecology and modern Chinese style. Our propose is learnt from Dinghu Landscape, Duanzhou Moyan and Southwest Guangdong culture etc, with repeat research and discussion, finally presented in exact forms with integration landscape. The completed Originality Office area is themed with "Zhesi Garden", featured with simple and abstract techniques in spatial division, unit facades, greening and landscape construction etc. It is designed to build a landscape environment of leisure in peace, creativity in thinking. Sun Cafe, decorative landscape walls, creative plant landscape doors, feature lighting wells, stone seats in natural shapes etc are positioned in the garden with sightseeing views around the whole garden but not crowded; it is fresh, sunny and friendly.
Another feature of the project is an artificial lake with an area of 20,000 m². It is set with Lakeshore Villas, Lake Island, Lakeshore Cycling Road, hydrophilic stairs, as well as

featured waterfront plants, native featured plants, presenting a lake garden with green and ecological landscape. The process of the project goes well so far, and we believe that we will work it out well. As designers, we are honor to take part in such a great comprehensive project and we will keep paying effort to improve our design and our service.

New House: How to interpret "Creation is an effective propellant between design and construction"?

Wu: Creation is mentioned because landscape design is not only applied in design; philosophically, it is a changing and moving object. The discrepancies between the design and construction happen in landscape projects, so we have to stress the importance of creation. I will further explained in 3 points:

Firstly, the on-site construction of landscape gardens is much more elastic than of architectures; it requires designers' on-site practice and the comparison between old and new techniques so as to match the age and the location. Currently, water-proof techniques are various and if a designer has no idea about the on-site situation, he will not able to know which way is the best.

Secondly, the construction of every project is varied with natural environments and artificial environment. Only the effective guidance of construction could improve the efficiency and save the materials. A great work could be accomplished by design based on the garden landscape features and style requirements, as well as the featured collocation; high-end materials are not indispensable. It is known that developers always figure out the ratio of input and output, with the expectation of less cost and more gains. My opinion is a good design with effective creation is a best way to lower the cost and waste in a project.

Finally, designers should use their sense of art to create and present the design completely in the final construction achievement, to guarantee the match of artificial buildings with the environment. It could be viewed practically in spatial transition, the selection of large-scale landscape stones, landscape arrangement, plants collocation and decoration etc. The interior decorative design could be sorted out as two kinds: the hard one and the soft one. The garden landscape should also be subdivided and the creation in the later stage should be paid more attention, which could enhance the importance of landscape design to a higher recognition level.

Only through the above three aspects and even more creation ways, could we well accomplish a project.

New House: How do you work with the construction side when following the project on-site?

Wu: Both of us have realized the necessary of creation in such a large-scale garden project---South China Wisdom Garden. So we had designers stayed in site and we also sent designers of different majors for guidance based on the project process, to guarantee the project's consistency and integrity, and the immediate handling and resolving of any problems. That's the advantage of design on-site. In fact, the site earthwork and buildings were carried on simultaneously, and the second, third and fourth phases are in the progress. Landscape construction could not be accomplished in one step; it is like playing a Go game, only far-sight and deep thinking could bring success. In the project of South China Wisdom Garden, landscape design is our main job and besides our requirement of the design quality, we also had to offer the reasonable and professional advices according to the requirement of the construction side. Take the revetment of artificial lake bottom for example, we presented several proposes for that and invited experts of ecological water treatment company and flood control for advices of the key problems. Besides, we held several conferences regularly with the material suppliers, purchasing department and marketing plan department etc, to guarantee the realization of the relationships before and after the project, and the commands from the construction side, so as to offer better guidance about the work.

New House: Compared with residential landscape design, what's the difference of big-scale industrial garden landscape design? What factors should be paid attention to?

Wu: Characteristics are based on the functions; the landscape design of large-scale industrial garden is on the basis of master plan construction requirement, themed with pieces sculptures to showcase the industrial garden's characteristics moderately with concise lines, pristine materials, the integration of leisure and ecological greening. Since Wisdom Garden is a high-tech industrial garden, it requires more wisdom flash, exploitation space and artistic inspiration inside. People work here could be relax, gain energy from the nature and enhance their work passion if the landscape design goes well. Landscape design is various but shares some principles, which are only different degree handling for different types of projects or different projects.

New House: Compared to landscape design of industrial park, what's the difference of large-scale specific industrial garden landscape design?

Wu: Tanshan Fuhua Heavy Industrial Park was one of my design projects, with a greening area of 200,000 m^2 and office buildings, factories, dormitories, pool clubs and VIP villas etc inside. This type of industrial park is open mainly to the people inside with greening and building a new-type industrial park as its goal, offering a good environment to its employees and a good image to its visitors. So its openness is relatively limited. Large-scale industrial garden like Wisdom Garden requires a featured landscape environment for itself and to attract more enterprises and groups, leading them to integration and coordination. This type of landscape is open, long-time and fluid, which requires designers' long-term follow-up and the team work with construction managers.

十年磨一剑：特色高端景观设计的实践者
—— 访上海唯美景观设计工程有限公司

■ **公司简介**

上海唯美景观设计工程有限公司成立于2002年，第一家分公司——武汉唯美景观设计工程有限公司于2005年在湖北武汉注册成立。唯美景观是一家专注于景观设计、高尔夫景观设计及相关工程的专业性公司，唯美设计机构包括唯美景观设计及圣安唯美高尔夫设计两部分。公司具有风景园林设计资质及工程资质，同时公司主创人员具有国际高尔夫场地管理资格证书，公司向客户提供从景观概念设计至施工图设计、现场施工指导及工程的全过程服务。主创设计团队，将高尔夫水系、地形设计理念完美融入景观设计中，结合城市规划与园林植物专业背景，将国际先进设计理念与中国地域文化完美结合，定制高端景观领地，营造户外梦想生活。

Company Profile:

Shanghai WEME Landscape was established in 2002, the first branch——Wuhan WEME Landscape was registered in 2005 in Wuhan, Hubei. WEME Landscape is a professional company that concentrates on landscape design, golf landscape design and related engineering, institutions of WEME Landscape including WEME Landscape Design and Shengan WEME Golf Design. The company owns landscape design qualification and engineering qualification, meanwhile the executive members possess international golf site management qualification certificate, provided customers with whole services that are from landscape concept design to construction drawing design, site construction guidance and engineering. Main design team integrated golf drainage, terrain design concept with landscape design, combined urban planning with their professional background of landscape plant, combined international advanced design idea with Chinese regional culture perfectly, created high-end landscape territory, built dream life outdoor.

《新楼盘》：在景观设计上面，你们倡导从概念到设计直至施工的全过程服务，这种"全过程"的服务对于景观设计有何重要的意义？

唯美景观：在景观设计上面，我们倡导从概念到设计直至施工的全过程服务，一般分为五个阶段：即总体概念方案规划设计阶段、方案深化设计阶段、扩初设计阶段、施工图设计阶段和施工配合阶段。

面对我国城市化进程的不断推进和市场开发程度的日渐深入这一形势，唯美景观提出了设计"全过程服务"的理念，在这一理念的指导下，唯美景观先后成功完成了多个重要项目。我司认为设计"全过程服务"是一种趋势，同时也体现出新形势下景观设计师所应具备的专业精神和社会责任感。

唯美景观根据自己多年从事景观设计经验，提出了设计"全过程服务"理念，这一理念认为设计师应该全程把控设计项目，而不应仅仅关注景观的设计本身，因为建筑、景观、规划这些设计都不是孤立存在的，它必然与周围环境息息相关。不仅如此，纸上的设计项目落地，也是一个复杂的过程。设计师如何保证自己的作品原汁原味的实体呈现出来，这也是设计师所应关心和考虑的问题。

"全过程"的服务景观设计是唯美景观过去十年设计实践的核心理念，这种景观设计认为一个景观设计师应该努力做到参与、影响并控制一个项目的全部过程。这种景观设计的表现是景观设计师在住宅、公共与商业项目开发中应该做到参与、影响与控制景观规划、景观设计、施工过程监理的全部过程。明朝的造园巨匠计成在他的造园专著《园冶》中提到"三分匠七分主人"的谚语，这里的"主人"是指主持造园的人，大体相当于设计者，也就是说造园要"三分施工七分设计"，指出设计之重要，设计所担负的责任之高。我们主张设计虽然是室内作业为主但也不是闭门造车。"全过程"服务是一个很好的方法，能更好地贯彻设计意图，更好地向实践学习，设计从实践中来，到实践中去。

唯美景观的理念是帮助客户成功，一个好作品的实际呈现，涉及到方方面面，包括投资方、设计单位、施工单位、监理单位等，"全过程"服务是其中非常关键的一环，也是实现各方多赢的重要步骤。

《新楼盘》：此外，唯美景观还秉承"精品、生态、经济、人文"的企业理念，那么应该如何去审视和理解这四者之间的关系呢？具体落实到项目上面又需要考虑哪些因素？

唯美景观：我们的企业理念是精品、生态、经济、人文。对精品的坚持与创造是唯美的精髓，我们追求精益求精的作品，源于设计理念的先进性，设计细节的完美性，设计手法的艺术性。基于这一企业理念而组建的优秀设计师队伍，科学的设计流程以及对设计作品的人文追求是唯美景观强大的核心竞争力。同时，景观生态在不同尺度上有不同的应用方式，生态群落、生态位、生态点、生态材料使人们能长久地从土地和环境中，源源不断地获得丰富感觉。以人为本，生态低碳的设计理念，通过最优化的设计与评估，善于利用各种不同价格的材料和当地材料，本地乡土树种，为业主着想，在预算许可的范围内达到景观效果最大化。精品、生态、经济、人文精神的完美结合造就梦想户外生活。

建筑师定义户内空间，景观师设计户外空间。唯美景观秉承精品、生态、人文、经济的企业理念，所有的一切功能设计是从人的使用出发。功能上要有人车分流、户外景观功能使用分区、人的行走与休闲等。比方说车行路就是从车的使用出发的，这是不同于以人为本的。另外，现在也开始提倡生态环境，要体现生物的多样性、低碳生活、建碳汇林、使用清洁能源等，这也是上海世博会留给我们的遗产之一。

唯美设计秉承精品、生态、人文、经济的企业理念，以"营造户外梦想生活"为唯美景观的使命，不仅在于提供完善的景观设计，而且还在于倡导人们全新的户外生活方式，一种与大自然亲切接触的理想户外生活。景观场所的建设，目的在于把人们引出户外，与大自然接触，与人接触，建立一种户外的交流之道与和谐的生活场所。在人们节假日及每天的闲暇时间里，来往并游憩于户外，令人心旷神怡。

《新楼盘》：在一些景观设计项目上，你们将高尔夫景观理念完美地运用于其他的景观设计中，这种灵感来自何处？对于项目的景观品质又起到了何种作用？

唯美景观：欧洲高尔夫场地教育的顶尖院校爱姆吾德学院(Elmwood College)是高尔夫的发源地苏格兰的唯一一所教育高尔夫场地管理学校。2006年初，它与上海交通大学(Shanghai Jiaotong University)合作，由圣安德鲁斯皇家古老高尔夫球会（The Royal and Ancient Golf Club of St. Andrews，简称The R&A）资助，在爱姆吾德学院培养高尔夫场地管理专家。经过选拔，唯美景观及圣安唯美高尔夫的负责人朱黎青先生得以进入爱姆吾德学院学习高尔夫场地管理专业，并获得由皇家圣安德鲁斯古老高尔夫球会认证之国际职业资格证书（Professional Development Award）。具有设计背景的PDA证书资格持有者中，朱黎青目前是全国为数不多的高尔夫设计师之一。同时作为景观设计专家和高尔夫设计专家，在全国更是少之又少。将高尔夫景观理念完美地运用于其他的景观设计中，这种灵感来自于长期的设计专业知识积累和设计实践感悟，非一朝一夕之功。

唯美景观将发源于苏格兰的自然高尔夫理念，与低碳及减碳观念结合，秉承环境友好理念，将高尔夫土地价值最大化的特性贯穿于住宅的景观设计中。同时，高尔夫舒缓起伏的地形、生态水处理、户外健康运动的精神都能很好的运用在我们的景观设计中；从而提升了景观设计的品质，营造唯美景观，成就户外梦想生活。

《新楼盘》：作为一家专业的景观设计公司，你们设计了众多优秀的景观项目，请简单介绍1～2个近期完成的项目，并谈谈其风格特点与设计特色。

唯美景观：我们迄今成立十年有余，已成功打造300多个景观项目。武汉未来科技城景观设计及世茂地产南昌世茂天城景观设计是我司最近完成的项目，下面着重介绍一下武汉未来科技城的景观设计：

快节奏，慢生活，打造科技未来新概念——武汉未来科技城景观设计。

武汉未来科技城位于武汉东南部东湖国家自主创新示范区，占地面积66.8 km²。其中，高新大道以北、外环线以西的2.6 km²区域为起步区，占地3800亩，规划有新能源研究院、研发区、孵化区、商务区及住宅区，其中起步区一期占地约500亩，在概念设计中称为A区。本项目即为A区范围内景观设计，该区容积率1.13，园区由A01～A09的九个地块项目组成。其中，原龙山水库排水明渠从园区中穿过，要求将排水明渠改造为龙山溪湿地景观公园，该湿地公园由不同标高的水面组成大小不同的湖面，分旱、雨季等不同的季节，形成"泉水叮咚"与"叠水"的效果，并配置相应植物，从而成为园区的亮点。项目地块即起步区一期（A区）位于东湖开发区东部，远离市区10 km。占地近33.1万m²，场地东侧为武汉市外环线，宗地北侧为龙山水库，自然景观优美。

科技园区的景观设计理念为："快工作、慢生活，打造科技未来新概念"；打造具有国际标准的、同时又具有本地特色的湿地景观与高科技园地环境。

景观设计遵循规划总体要求：体现生态性、开放性、先进性、科学性四项原则，突出"共生、绿色、开放、人文"四大理念。从地面景观到屋顶绿化，打造立体的城市绿化系统，极力打造——生态、和谐之城；绿色、低碳之城；开放、共享之城；人文、创新之城。

项目特色：创新性——在设计与实施过程中，应加强高新技术、新材料、新工艺的运用；建筑及小品——仿生建筑的艺术设计；第五立面——屋顶花园的生态运用。

《新楼盘》：今年是唯美景观成立的第十个年头，可谓"十年磨一剑"，这十年给你们最大的感悟（收获）是什么？对于未来，唯美景观又将有怎样的美好憧憬与规划？

唯美景观：今年是唯美景观成立的第十个年头，可谓"十年磨一剑"，回首这十年，我们充满自豪，展望未来，我们满怀信心。

我们公司的目标是在未来三到五年内打造成国内优秀的专业景观设计公司及地产商的最优选择之一。通过十年、三百多个项目的积累，我们已经进入了主流的景观设计公司行列。下一步我们要更着重精品的打造，注重生态、经济、人文的理念，在生活中构筑自然，为客户提供诗意盎然的唯美景观。我们的市场定位，是特色高端景观设计服务提供商，我们的设计范围涵盖城市设计、高尔夫规划设计、景观规划设计三个类型，而且我们有亲手完成这些工程项目的能力。

目前国民经济的高速增长，有一定的粗放性，它对于环境有日益增加的欠债。在可以预计的时间内，这种状态还会持续。景观设计行业，在这一经济发展的进程中，将会大有作为。按目前城市化增长速度，20年后我国的城市化到达70%，这一过程按目前每年1%的城市化增量。20年以后，可能城市建设的速度会大大放缓，而景观还将持续相当长一段时间。

官方估计，2015年中国人口会到顶峰然后开始下降；随着水稻种植科技的提高，未来恐怕18亿亩耕地的红线会下降，也许只需15-16亿亩？这会释放出大量的地用于建设。

未来20年会怎么样呢？人口总数绝对下降；更多的城市人口（超过70%，甚至更高），更少的农村人口；粮食种植科技的进一步提升；更多依赖清洁能源；对森林碳汇的诉求更加强烈；从生存到享受到进一步享乐的转变；不到20年的时候，中国将是世界第一大经济体，从静态的角度计算人均收入也会到20 000美金的程度，相当于超过现在的台湾地区，达到葡萄牙的人均富裕程度。如果从购买力的角度，则会比肩当今七国集团。当然到那个时候，中国的建设速度会绝对降低，但我们还是非常看好景观设计在未来十年、二十年的发展。

INTERVIEW | 专访

A Decade of Practice for Characteristic and High-end Landscape Design
—— Interview With Shanghai WEME Landscape

New House: On landscape design, you advocate whole services form concept to design till construction, what does this kind of 'whole' services mean to landscape design?

WEME Landscape: on landscape design, we advocate whole services form concept to design till construction, which are divided into five stages generally: namely overall concept program planning design phase, deepening plan design stage, expanding initial design stage and construction drawing design stage and cooperative construction stage.

Confront with the tendency of our country's continuously urbanization advancement and gradual deepening of market development, WEME Landscape put forward the design concept "whole service", under the guidance of this concept, WEME Landscape has successfully completed several important project. We think that "whole service" design is a kind of trend, which also reflect that the landscape architect should have of professional spirit and the sense of social responsibility under new situation.

According to several years' design experience, WEME Landscape put forward the design concept of "whole process service", the concept acquired that the designer control project all the way, and should not only focus on landscape design itself, because the design of architecture, landscape, planning are not isolated from each other, they related with surrounding environment closely. More than that, when the projects on paper carried out, the process is also a complex process. How the designer guarantee their works be presented with authentic entity, which is also a question that the designer should consider.

"Whole process" service landscape design is the core idea of WEME Landscape during the past decades, it is thought that a landscape architect should work hard to participate, influent or control the overall process of a project. The performances of landscape design are that landscape architect should be involved in, influence and control the landscape planning in the development of housing, public and commercial projects, landscape design and construction process of the supervision of the entire process. Landscape master Jicheng in Ming dynasty mentioned the proverb "three points fitter seven points master" in his landscape monograph Yuanzhi the, here the "master" refers to the people who built the yard, roughly equivalent to the designer, that is to say landscape design needs "three points construction seven points design", which pointed out the importance of design and the high responsibilities a design institute should bear. We advocate that though design is interior activity which should be given priority to but not work behind closed doors.

"Whole process" service is a very good method, it made implement the design intention better, learn from practice better, design should from practice and into practice.

The concept of WEME Landscape is to help the customer get success, a real presentation of a good work, which involves many aspects, including investor, design unit, construction unit, supervision unit, etc, "whole process" service is one of the important parts of the process, also an important step to achieve multi-win of all parties.

New House: In addition, WEME Landscape still sticks to the corporate philosophy of "boutique, ecological, economic and humane", and how to examine and understand the relationship among the four factors? When putting into practice, what factors need to be considered?

WEME Landscape: Our corporate philosophy is boutique, ecological, economic and human. To persist on and create boutique is the essence of aestheticism. We pursue the excellence of the works, the advanced nature of design concept, the perfect of details and the artistic quality of the design skill. Based on the enterprise idea, the good designer team, scientific design process and humanistic pursuit of the design work compose the strong core competitiveness of WEME Landscape. At the same time, the ecological landscape is used in different ways, such as ecological community, ecological site, ecological point, ecological materials that people can constantly get rich feeling from the land and environment. People-oriented is the design concept of ecological low carbon. By designing and evaluating optimized, fully using of all different prices of materials and local materials and taking the owners' profit into consideration, the landscape effect can be maximized to the largest extent in the permitted scale. The perfect combination of boutique, ecological, economic and human spirit creates a dream outdoor life.

The architect defines the indoor space, while the landscape designer takes charge of the outdoor space. WEME Landscape sticks to corporate philosophy of boutique, ecological, economic and human, and starts all functional design from people's need. Functionally, there are people and cars apart, landscape functionally zoning, people walkway and leisure, etc. For example, garage road is used for cars, which is different from the people-oriented. In addition, now we also began to advocate the ecological environment, focusing on the biological diversity, low carbon life, carbon collecting forest, and clean energy use and so on, which is one of the heritages Shanghai world expo left us.

WEME Landscape sticks to corporate philosophy of boutique, ecological, economic and human, keeping "create dream outdoor life" as the WEMW Landscape mission. We not only to provide perfect landscape design, but also lead people to a new outdoor lifestyle which is an ideal outdoor life connecting tightly with nature. The construction of landscape place is to encourage people going outsides to enjoy nature and contact with other people, setting up a kind of outdoor communication way and a harmony place to live. In people's holiday and idle time, dealings and having a rest in the outdoors makes a person more relaxed and happy.

New House: In some landscape design project, you perfectly place golf landscape concept in other landscape design, and where the inspiration comes from? For the landscape quality of the project, what kind of role it plays?

WEME Landscape: Elmwood College is the only golf education management school in Scotland, the birthplace of golf. At The beginning of 2006, it had cooperation with Shanghai Jiaotong University and was subsidized by The Royal and Ancient Golf Club of St. Andrews, (referred to as The R&A), training Golf management experts. After selection, WEME Landscape and the principle man, Zhu Liqing qualified to enter Elmwood College to take the golf site management course, and receive the Professional Development Award. Among the PDA qualification certificate holders, Zhu Liqing is the one of the few golf designers at present. At the same time, it is rare to being a landscape design experts and golf design experts in the whole nation. The inspiration of perfectly placing the golf landscape concept in other landscape design comes from the long-term design professional knowledge accumulation and practical design feeling, not in one day of work.

WEME Landscape combines the Scotland natural golf idea with low carbon and carbon reduction concept and adheres to the environment friendly concept to maximize the golf land value throughout the residential landscape design. At the same time, the slow undulating golf terrain, ecological water treatment and outdoor health sports spirit can be used well in our landscape design; thus promoting the quality of the landscape design, creating a beautiful landscape and achieving the dream outdoor life.

New House: As a professional landscape firm, you have designed a lot of great works. How about introducing one or two of your recent design projects? What are their features and styles?

WEME Landscape: We have established more than ten years so far, and have been successful to create more than 300 landscape projects. Wuhan Future City and Nanchang Shimao Skyscrapers are the latest two works of our firm and I would like to share some details on Future City with you.

Fast-paced, living slowly, creating technological future new concept – landscape design of Wuhan Future City.

Located in Wuhan southeast East Lake national independent innovation demonstration zone, Wuhan Future City covers an area of 66.8 square kilometers. To the north of Gaoxin Avenue and to the west of out ring road is the initial area that covers about 2.5 square kilometers, including new energy research institute, the research and development area, the hatch area, business district and a residential area, in which the first phase is about 0.33 square kilometers, known as Area A. Our design is just for this Area. With a plot ratio of 1.13, it consists of nine plots from A01 to A09. Original Longshan reservoir drainage runs across the area, which was then transformed into Longshanxi Wetland Landscape Park. Lake surfaces in different sizes and levels in the park produce different visual effect and audible beauty in different seasons, decorated with appropriate plants and thus become the highlights of the park.

Philosophy of this project is: Fast-paced, living slowly, creating technological future new concept; create international standard wetland landscape and high-tech garden environment with local characteristics.

The landscape design follows the planning overall requirements: embodying the four principles about ecology, openness, advancement and scientificalness, highlighting four concepts of symbiosis, green, openness and humanities. To create a three-dimensional landscape from the ground to the roof greening, to build a city of ecology and harmony, city of green and low-carbon, city of openness and sharing and city of humanities and innovation!

Project Characteristics: innovativeness—strengthening high and new technology, new materials and new technique; architecture and featured landscape—artistic designing of bionical architecture; fifth facade—ecological roof garden.

New House: This year is the tenth year of your firm. What was your greatest harvest in the past decade? And what's your vision in the future?

WEME Landscape: Look back on the past decade, we are a proud firm. Look ahead, we are full of confidence.

Our goal is to be an excellent and professional domestic landscape design company and one of the optimal choices for real estate agents in the next three or five years. With more than 300 projects in the past decade, we have entered the ranks of the mainstream landscape design firms. Next we want to build more quality products, focusing on the concept of ecology, economy and culture, building nature in daily life, to provide customers with poetic WEME landscape. We aimed to be a high-end landscape design service provider. Our design covers three types such as urban design, golf planning design and landscape design, and we have the ability to complete these projects.

At present, the rapid growth of national economy has a certain ruggedness, which owes an increasing debt for the environment, and this state will continue in estimated time. Landscape design industry will make a difference in this process. According to the current 1% growth rate of urbanization, China's urbanization ratio will reach 70% 20 years later. At that time, maybe the urban construction will slow down considerably, but the landscape will continue for quite a long time.

Official estimated that China's population peak will begin to decline by 2015; with the improvement of rice cultivation technology, the red line of the 1.8 billion hectares of arable land will decline, perhaps just 1.5-1.6 billion acres are needed. This will reserve a lot of land for construction.

What will happen in the next 20 years? Total population will decline absolutely, there will be more urban population (more than 70% or even higher) and less rural population; further enhancement of grain cultivation technology; greater reliance on clean energy; more intense demand for forest carbon sinks; living for enjoyment even further; in less than 20 years, China will be the largest economic entity in the world, per capita income will be 20,000 dollars from a static point of view. By that time, China's construction speed will definitely be reduced, but we are still very bullish about landscape design in the next decade or two decades.

ELEGANT AND SOPHISTICATED NEOCLASSICAL CULTURAL LANDSCAPE

| Yangzhou Zhongji Zijin Wenchang Landscape Design

古典优雅 精致从容的新古典主义人文景观
—— 扬州•中集紫金文昌景观设计

项目地点：中国江苏省扬州市	**Location:** Yangzhou City, Jiangsu Province, China
开 发 商：扬州市中集达宇置业有限公司	**Developer:** Yangzhou Zhongji Dayu Properties Limited
景观设计：深圳市雅蓝图景观工程设计有限公司	**Landscape Design:** Shenzhen Yalantu Landscape Engineering Co., Ltd.

　　本项目位于扬州市维扬区东西中轴线文昌路中心位置，东至富达路与市政府一路相隔，南至文昌中路对面为街心公园，西邻海关大厦至维扬路，北至顺水大酒店商界一线。规划总用地约50 000 m²，总建筑面积约160 000 m²。该项目包括高尚居住社区、休闲会所、5A级金融商务办公楼等。设计使用了新技术、新工艺，引入了"智能化、绿色环保"元素，充分考虑了与周边建筑物的协调及城市的互动，采用双首层、双大堂、空中花园、屋顶花园等设计手法，将形成立体的、多层次的生态、绿色、健康空间。

　　本方案设计秉持中集集团的"自强不息，挑战极限"之道，以尊贵与精致为景观设计的主要思想，以新古典主义景观风格为设计理念，遵循"文化、精致、优雅、自然、私密"的设计原则，臻于每一处点滴的创作与营造。根据对本项目的分析，景观设计主要是以新古典主义风格及主题，极力打造一个全新的"古典优雅，精致从容"的商业办公及居住的豪华高档的人文景观环境。

商业办公空间

　　商业办公空间通过对建筑风格的充分把握，在铺装形式材料、水景、灯柱、雕塑小品、商业LOGO墙、伞架椅、城市的景观空间划分及城市的标识物等延伸大气沉稳的现代建筑形象；城市商业广场要求通透性，以展示其时尚高贵大气的一面，植物以点、线状为主，不宜过多，同时会以本土树种为主，合理布置珍稀植物树种来营造异域风情；居住空间结合项目的规划布局，景观设计主要分为五大区域构成：中心花园区、运动休闲区、会所景观区、主入口景观区、商业区和别墅区等来表达新古典主义风情园林景观。

总平面图 Site Plan

住宅空间

1. 豪华大气的主入口广场，简明的花圃围合社区标志墙，跌水与景观置石将更多的空间留给了人们的视线而不是脚步；

2. 大片绒毯般柔软的庆典草坪，散发泥土和青草混合的芬芳，如同屏幕般的高大乔木，隔绝喧哗，尽可体味别样的静谧；

3. 专为孩子们准备的孩趣天地，给孩子们充分撒野、玩耍、锻炼的机会，享受恣意的童年时光；

4. 通过铺装模块中加入自然、随机的种植空间，以及巧妙的与建筑空间相结合等手法，使人们不约而同的因美景而聚合在一起，欢享生活的情趣盎然；

5. 采光井引入时尚的设计理念，将采光井的设计与景观设计融为一体，设计注重良好的通风和良好的自然采光，符合当前国际提倡的低碳环保理念；

6. 设计以满足消防车道要求为主，以求提高住宅区内的居住舒适环境和安全环境，营造一个大气高档豪华住宅区；

7. 住宅区围墙内到建筑的周边一带为噪音缓冲区，以浓密种植为主，减少外界噪音对居住区的干扰，提高区内安静的居住环境。

The project is located on Wenchang Road, Weiyang District of Yangzhou, the Fuda Road in the east with the municipal government here, south of Wenchang Road opposite the boulevards Park, west of the Customs Building to Weiyang Road north to sailing Hotel business community first-line. Planning a total land area of about 50,000 m^2, total construction area of about 160,000 m^2. The project consists of the noble residential communities, leisure clubs, 5A-class financial and business office. Design using new technologies, new processes, and the introduction of "intelligent, green" elements to take full account of the interaction and coordination of the surrounding buildings and cities, using a double first floor, the double lobby, sky garden, roof garden design practices will form a three-dimensional, multi-level ecological, green, healthy space.

This program designed to uphold the CIMC "self-reliance, to challenge the limits" spirit, with noble

NEW LANDSCAPE | 新景观

and exquisite landscape design of the main ideas, neo-classical landscape style design concept, follow the "culture, sophistication, elegance, nature, privacy design principles, are reaching every bit of creation and construction. Landscape design is mainly based on the neo-classical style and theme based on the analysis of the project, trying to build a new "classic elegance, sophistication and calm" commercial office and residence luxury high-end cultural landscape environment.

Commercial and Office Space

Commercial office space through a full grasp of the architectural style, LOGO wall, paving materials, water features, lamp posts, sculptures, sketches, commercial umbrella chair, city landscape space is divided city markers such as extending the calm atmosphere of modern architecture the image; city commercial plaza requirements permeability to showcase their stylish

elegance and atmosphere, plants to the point, the linear main, not too much, mainly native trees, rational arrangement of rare plant species to create exotic; living space combined with project planning and layout, landscape design is divided into five regions form: central garden area, sports and leisure clubs landscape area, the main entrance of the landscape area, business district and villa area to express the neo-classical style landscape.

Residential Space

1. Luxurious main entrance plaza, concise flower garden Community walls, water or landscape stone leaving more space for sight of the people rather than the pace of people;
2. Large areas of carpet celebration soft lawn, dissemination of dirt and grass mixed fragrance, like a screen like the tall trees, isolated noise as much as possible to appreciate a different kind of quiet;
3. Prepare the children for the children interesting world, fully to the kids to run wild, playing, exercise opportunities, enjoy willful childhood;
4. Pavement module added natural the random planting space, and clever architectural space combined approach, so that people invariably beauty coming together to celebrate a fun-filled life;
5. The light wells introduction of fashion design concept, the integration of design and landscape design of the light wells, designed to focus on the good ventilation and natural lighting, low-carbon environmental protection concept in line with the current international advocate;
6. Designed to meet the fire road requirements mainly in order to increase the comfortable living environment and security environment of the residential area, creating an atmospheric luxury residential area;
7. A noise buffer to the immediate area of the building in a residential area, dense planting to reduce the interference of outside noise for residential areas and improve regional and quiet living environment.

项目地点：中国广东省阳江市阳东县
景观设计：广州市圆美环境艺术设计有限公司

FRESH AND PASTEL LIFE-PRESERVING GARDEN LANDSCAPE WITH ELEGANT REALM | Yangjiang Danmo Cabin

清新淡雅、意境优美的养生庭院景观——阳江淡墨幽居

项目地点：中国广东省阳江市阳东县
景观设计：广州市圆美环境艺术设计有限公司
庭院营造师：吴应忠

Location: Yangdong County, Yangjiang, Guangdong, China
Landscape Design: Guangzhou Yuanmei Environmental Art Design Co.,Ltd
Designer: Yingzhong Wu

本项目位于阳江市阳东县城新华路以北、龙塘路以东，此次设计为中惠沁林别墅豪宅养生庭院——"淡墨幽居"的景观设计部分，占地面积2 200 m²，庭院景观设计面积1 600 m²。项目所在的中惠沁林山庄总占地面积18万m²，总建筑面积30万m²，产品定位是"引领粤西名门生活"，项目定位是"引领阳江豪宅生活"，在阳江市打造一个低密度、高品质、高档次的复合型社区，类型有花庭美墅、联体别墅和八栋高层。

设计主题：以"远山近水群岛若画里，碧叶朱花小草皆有情"为构景意境，采用了水墨为主元素，墨分五色"干、湿、浓、淡、焦"，一管狼毫，一块端砚，人事百态，人生起伏，尽现纸上。墨代表行为着迹，与景观设计中的疏密、轻重、主次布景相联系；水引自"上善若水，水善利万物而不争。"之语，所谓最高境界的善行就像水的品性一样，泽被万物而不争名利，水代表德，心的体验，无形而至高，恰似信步闲庭间，静坐庭院一隅，品一缕茶香，静思沉酿。

整个庭院设计希望将精神层面的内容赋予景点中去，让景观具有生命力，同时这种精神是最大地接近业主的价值观，以达到业主在起居之时，能感受到庭院与自己的对话，共呼吸，汲取自然的力量，释放自我，启迪思想。

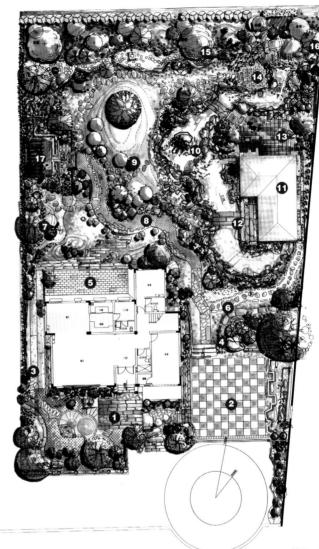

① 客似云来
② 六合阴晴
③ 红叶寄情
④ 淡墨幽香
⑤ 万籁无声
⑥ 归来悟空
⑦ 笑傲风月
⑧ 蛟龙得水
⑨ 风轻云淡
⑩ 沧浪之水
⑪ 墨趣轩
⑫ 双源桥
⑬ 抚琴台
⑭ 竹林风韵
⑮ 观山揽翠
⑯ 飞珠溅玉
⑰ 归隐田园

总平面图 Site Plan

NEW LANDSCAPE | 新景观

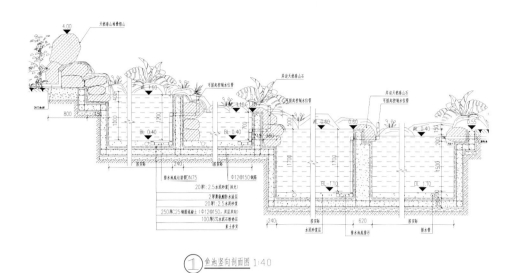

① 鱼池竖向剖面图 1:40

This project is located in northern Xinhua Road, eastern Longtan Road in YangDong county, Yangjiang City, the design is the landscape part of Danmo Cabin which is belong to Huiqinlin villa life-preserving garden, covers an area of 2,200 m², and garden landscape design area of 1,600 m². Huiqinlin Villa covers an area of 180, 000 m², with a total floor area of 300, 000 m², the villa positioning is "leading the noble life in western Guangdong", and the project positioning is "leading luxury mansion life in Yangjiang", build a low density, high quality and high grade complex villa community in Yangjiang, the type including garden villa, townhouse and eight high-rises.

Design theme: with "the beautiful scenery of mountain, river, and island is like a painting, green

leaf, red flower and grass as if have emotion" as landscape artistic conception, using ink as its main elements, the ink is divided into five colors: "dry, wet, strong, weak, coke," a Chinese writing brush, a piece of Duan ink, life phenomena, ups and downs, are showed on the paper. Ink presents trace, which is connect to density, weight, the primary and secondary scenery of and landscape design, water originates from the famous quotation "as good as water, water is good at everything while indisputable", the highest level of good deeds is like the character of water, spread all-round benefit to all but never ask for fame and wealth, water stands for virtue, the experience of heart, invisible but supreme, it seems to walk around an idle court, sit at the corner of the courtyard, taste a cup of tea, meditate for a while.

The designer hope to give some spiritual things to the landscape design, and make the landscape full of vitality, meanwhile, the spirit should be close to clients' value, to achieve such a result when the owners living here, they can feel the yard has a conversation with themselves, they can draw power from nature, release themselves, open and enlighten their thinking.

FEATURE | 专题

年度最佳楼盘与景观

专题导语

随着年末将至，各种年度的楼盘评选活动及年度的优秀作品也逐渐揭开面纱。回顾即将过去的这一年，《新楼盘》杂志专题紧紧抓住了"图解地产与设计"这条主线，为广大的读者呈现了更加丰富多彩的专题内容。纵观整个年度的专题项目，各种类型新颖的楼盘与景观层出不穷，各种新技术、新材料、新理念的应用也让楼盘设计更趋合理，设计专业化与细分化趋势进一步明显……本期专题，将从众多的优秀项目中挑选出部分年度最佳楼盘与景观的代表作品，作为一个年度的盘点，希望这些作品能够代表过去一年里各类建筑与景观的设计趋势和发展方向，我们将与大家一起分享。

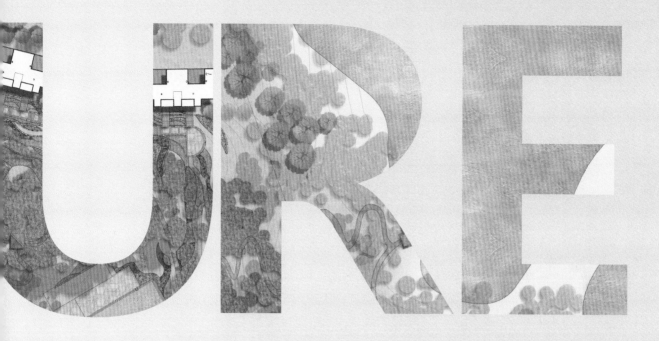

Introduction

As we approach the end the year, various upcoming annual house selection will happen and the best works are about to be released. Look back on the past year, New House had firmly grasped the main theme of illustrating property and design, presenting more and more colorful topic content for readers. View the special works of the whole year, new buildings and landscapes in various types emerge in an endless stream; new technologies, new materials and new ideas improve more reasonable designs; the trend of design specialization and segmentation is more apparent······In this special topic, we select some representative building and landscape works from numerous outstanding projects. As an annual check, these works would be able to represent the design tendency and development of building and landscape in the past year. We hope you enjoy.

和丰置业（佛山）有限公司
HEFENG PROPERTY CO. LTD

企业介绍
和丰置业（佛山）有限公司成立于2006年4月，是一家以房地产开发为主的全外资企业，涉及工业地产、商业地产和住区住宅三大类别，同事从事相关的房地产销售、租赁及物业管理等业务。

产品与服务
和丰置业秉承"社会效益、环境效益和企业效益最大化，客户满意最大化，员工个人发展空间最大化"的经营理念，专注于精品房地产产品的开发，积极参与所在区域的城市化建设，强势打造具有时代特征、品牌意识、创新精神和高性价比的精品项目。自公司成立以来，先后为各类业主提供了大批优质的地产项目，以其良好的服务意识赢得了广大客户的认可。

Company Profile
Hefeng Property Co. Ltd was founded in April, 2006, which is a wholly foreign-Owned enterprise being engaged on the development of real estate, involving industrial real estate, commercial property and residential housing. In the meanwhile, it also proceeds property related business like selling, leasing and management of real estate properties.

Product and Service
"To maximize the social benefits, environmental benefits and enterprise benefits, and to maximize customer satisfaction and the space for employees' personal career development" is the operation philosophy that Hefeng Property adheres to. Hefeng Property focuses on the development of high-quality property products, actively take part in the urbanization where it belongs to, strong build fine projects with times feature, brand awareness, creativity spirit and high cost-effective. Since from Hefeng's foundation, abundant high-quality property projects have been provided to the owners one after another; and Hefeng earns the recognition from customers for its good service awareness.

江西天沐置业有限公司
TIANMU GROUP

企业介绍

天沐集团是由天沐温泉旅游投资集团公司和天沐地产开发集团公司组成的一个集投资、开发、销售、经营管理一体化的复合型产业集团。

产品与服务

天沐地产开发集团——在天沐温泉多年打造成形的度假区基础上发展起来的地产专业开发公司。公司充分利用休闲的人居环境，借助知名旅游景区优越的自然条件，着力开发高端别墅、度假公寓、产权酒店、花园洋房、企业公馆等多类型产品。天沐地产遵循"保护自然、亲近自然和享受自然"的开发理念，努力为成功人士打造温泉养生、休闲度假、投资置业的高端物业。公司在明月山开发的"天沐温泉谷"、庐山开发的"天沐美庐荟"依靠完美的品质在业内和客户间树立了良好的口碑。

Company Profile

Tianmu Group is a compound industry group integrating investment, research and development, sales, and management, It is jointly set up by Tianmu Hot Spring Resort Group Co. Ltd and Tianmu Real Estate Development Co.Ltd.

Product and Service

Tianmu Real Estate Group – a professional company engaging property development. It fully uses the leisure settlement and draw support from the advantageous natural conditions of famous tourism areas, to develop diversified products, such as high-end villa, vacation rentals, property hotels, garden houses and cotels. "To protect nature, get close to nature and enjoy the nature" is the developing philosophy Tianmu Real Estate adheres to, endeavoring to build up high-end properties that possesses hot spring spa, leisure resort and could be invested. Right now through perfect qualities, "Tianmu Hot Spring Valley" developed at Moon Mountain and "Tianmu Mei Lu Hui" developed at Lu Mountain establish a good reputation among this industry and customers.

浅谈楼盘景观主题化设计要点

■ 人物简介

陈英梅
2000年毕业于华南农业大学
曾任美国AAM集团公司（广州景观部）设计主管
现任广州市圆美环境艺术设计有限公司 设计创意总监
擅长在景观设计中融入主题和文化，追求令设计更富于内涵和生命力，想象无限，创意无处不在。

Profile:

Chen Yingmei
Graduated from South China Agricultural University
Ever worked as the design supervisor of landscape department (Guangzhou) for AAM Group
At present, the design director of Guangzhou Yuanmei Design
Be skilled in combining theme and culture into landscape design with limitless imagination and creativeness.

上个世纪90年代末，广州的十大名盘翠湖山庄在景观方面以突出"万象翠园"为主题，把欧式、美式、日式、中式景观元素融汇进各个展示空间，以集锦式园林造景手法，体现景观风格的多样性，其创造性均对广州房地产开发产生了颇为深远的影响，引致后来者纷纷效仿，并使广州地区的楼盘景观进入一个越来越备受重视的时段。

楼盘具有鲜明的主题，是基于其本身楼盘的环境格调、建设规模、建筑特色、营销需要的，而成功的景观反过来会大大提升楼盘的整体质素。在个性化需求更高的未来，主题化景观设计将是一个潮流趋势。下面以笔者主设的三个楼盘项目来进行探讨。

1、一个楼盘的景观，如果确立了一个明确的主题，那么它将贯穿景观整体布局到细节刻画中。笔者设计的一个项目，位于广州从化区，主题为"音符上的四季"，以富有特色的奥地利音乐文化为蓝本，利用园林的造景手法，营造一个充满音乐元素，艺术灵感浮现，又兼具欧式文化特征，雅致、灵动、高品位的小区环境。平面构图上，采用了多个五线谱音符构图，组成主园路、儿童游乐、休憩花圃、水景等空间，而花架、跌水级、流水构件等单体也采用了钢琴、竖琴等元素来刻画，将主题的形式化达到更大的表现。

2、某些景观的主题化更多的表现在内涵上，它没有大面积的图形勾勒，需要在单体构造、空间层次、文化小品上进行渲染设计的创意。云浮新兴的御龙庭国际公寓，在其建设范围内，对着市政路交叉口，有一座"二龙亭"，在当地具有一定的历史，故需要对其进行修缮保护。所以在设计上，我们以该亭为景观原点，将其周围场地发散设计为一个文化小广场，利用通透景墙、砂岩小品、篆字景石等组合成一个富有文化气息的开放空间，很好地提升了楼盘气质。

3、景观主题化设计要与实际相结合。在增城百花大道上的一个别墅楼盘"泊爵"，原有的场地是一片平地。在开始与甲方沟通的过程中，甲方一直有想法去建设一个东方威尼斯的景观，希望有一个水系环绕整个楼盘，但是这样的建设势必要以业主交纳高额管理费来实现四时流水的景观。

后来在和规划部门的商谈中，我们了解到在项目的北边有一天然水库，于是提出引此水作为小区内景观用水的方案。行政方面的工作交由甲方处理，我们负责翻查了多个技术资料，对水库水的引入，和引入后与各景观点的给排水衔接控制，如水上车行道、流水挡土墙、排洪等等进行了研究，并寻找专业公司进行技术支持配合，最终实现建设成一个游船徜徉、水鱼旖旎的景观。解决了水的问题，再回到实质设计，首先由于威尼斯处于地中海，其多数的植物品种是不能在增城种植的，这一点决定了不能做一个完全的威尼斯景观；其次楼盘的建筑形式也非纯粹欧式，采用的材质为高档石材与质朴石料、涂料等结合；再次作为别墅楼盘，各栋别墅的花园建设势必会呈现百花齐放的姿态，风格不可能做到统一化。综合考虑后，我们为该楼盘定义了一个融东方威尼斯水系，与热带东南亚风情结合的景观，希望通过小运河的大环境体现优雅气质，茂盛的植物表达蓬勃生机，混搭的色彩渲染出艺术的无限可能。

With Concise Remarks on Landscape Theme Design for Housing Projects

In the end of 1990s, Lakeshore Villa – one of the top ten Guangzhou housing projects was famous for its theme landscape "Panoptic Green Garden" which combines the landscape elements of European style, American style, Japanese style and Chinese style into different spaces. It shows diversified landscape styles in one garden and influences the real estate developments of Guangzhou for a long time. Many projects followed its style and the landscape design for housing projects drawn more attention from then on.

The theme of a housing project will decide the landscape style, project size, architectural style and its marketing. And a successful landscape design will greatly improve the quality of the whole project. With the higher demand for identity, designing landscape with distinct theme will be a trend. Here, three projects designed by myself will be used to make my remarks on this topic.

1. Once decided, the landscape theme of a housing project will penetrate into every detail. One of my works, located in Conghua District of Guangzhou City, is designed with the theme – "four seasons on musical notes". Distinctive Austrian music culture was introduced to create an elegant and high-taste environment with music elements and artistic inspirations as well as European culture. In terms of site plan, the musical note compositions are applied to form the garden paths, children's playground, leisure flower beds, waterscapes, etc. Whilst the flower stands, cascades, running water, etc. are shown in the form of piano or harp. Thus the theme was emphasized again.

2. Some landscape theme shows in the contents but not in the large area or scale. It needs creative space design, landscape architecture and cultural elements to highlight the theme. For the project Yu Long Ting in Yunfu City, there is the "Double-dragon Pavilion" facing to the road intersection. It is a relic need to be renovated and protected. Therefore, we take this pavilion as the original point and around it a small cultural square is designed. Landscape wall, artificial hill and landscape stones are used to create an open cultural space for a high quality development.

3. The landscape theme should consider the surrounding environment. The site of "Earl Villa" on Baihua Avenue of Zengcheng is a flat land. During the communication period, the client wanted to create an Oriental Venice with a river running around the project. However, it must cost a lot in maintainance and management for this effect. Afterwards, when talking with the planning department, we got to know that there is a reservoir on the north of the site. The proposal to make use of this facility thus came into being. So the client began to apply for a permission and we did many researches on the use of water from reservoir and the drainage system. And then we found the technical support from professional company and finally realized the goal. After solving the water problem, we came back to the detail design. Since Venice locates in Mediterranean area, most of the plants there cannot grow in Zengcheng. In addition, the buildings are not in pure European style but designed with high-end stones, common stones and paint. What's more, the garden design for different villas will be different in style. Therefore, we decided to create a landscape system which combines the water system of Oriental Venice with South Asian landscapes. We hope to provide an elegant and artistic environment with small canal, luxuriant plants and mixed colors.

IMITATING NATURE, LOW-PROFILE LUXURIOUS EXQUISITE VILLA COMMUNITY | Lake Side

师法自然、低调奢华的精致别墅社区 —— 鹭湖

项目地点：中国河北省香河县
开 发 商：廊坊旷世基业房地产开发有限公司
　　　　　廊坊万恒盛业房地产开发有限公司
建筑设计：北京三磊建筑设计有限公司
建筑面积：225 308 m²

Location: Xianghe County, Heibei Province, China
Developer: Langfang Kuangshi Jiye Real Estate Co., Ltd
Langfang Wanheng Shengye Real Estate Co., Ltd
Architectural Design: Beijing Sunlay Design
Area: 225, 308 m²

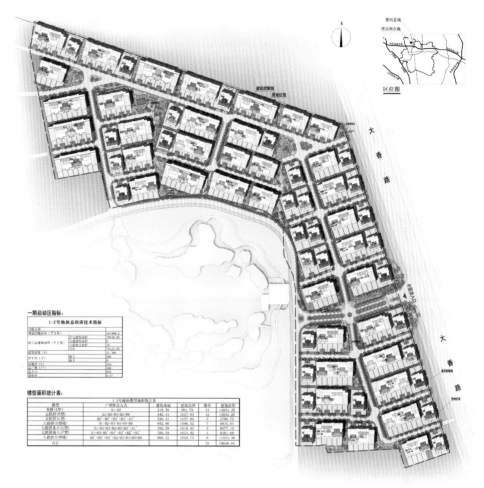

总平面图 Site Plan

项目概况

项目位于香河县中心区西北部，地块东侧紧临城市主干道大香路，现状蒲池河流经地块南侧。建筑面积为225 308 m²，其产品以低密度的联排别墅和双拼别墅为主。

规划布局

项目秉承师法自然，与环境相融合的设计理念，通过对场地周边环境的研究解读，明确以大地肌理作为规划设计模板，形成丰富的规划肌理，同时结合用地进行的住宅布置使得多余的边角用地成为公共景观绿地，为居住者提供了交流、游憩的场所，提升了住区的整体居住生活品质。

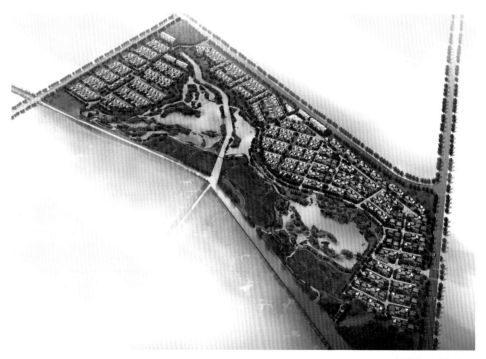

鸟瞰图 Aerial View

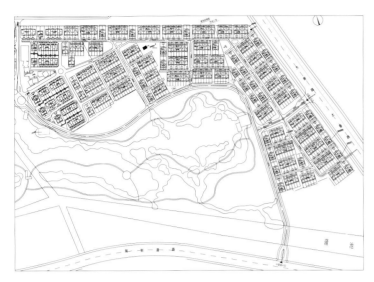

交通分析图 Traffic Drawing

绿化分析图 Green Analysis Drawing

建筑设计

借鉴荷兰现代建筑风格，以灰色体块为基底，白色阳台为横向联系；灰白色系体块上搁置不同材质，不同颜色的45度坡屋顶小屋，构成了别墅区整体淡雅清新的基调。本色质朴的红色陶瓦、木瓦、以及竖向线状肌理的陶板、面砖等外墙材料的穿插使用，为庄重典雅的灰白基调带来些许的活跃与惊喜，并将三层高的建筑分解，使之降为宜人尺度。阳台扶手面与线之构成，灰色劈开砖的铺砌变化，白色仿石漆墙体的深色压顶，直立锁边系统铝板的封口，玻璃天窗的处理等细节设计，令整个建筑简洁而精致，体现追求低调而高品质的设计目标。

景观设计

通过建筑的错动和建筑外立面色彩、材质的变化，布置形成富有情趣的、变化的公共景观空间并穿插于街巷中，构筑起该片区的重要景观轴线和节点，同时也成为邻里公共生活空间和重要交流平台。效仿彼埃·蒙德里安的抽象几何绘画，对标准化设计的住宅原型进行排列组合，住宅间形成的空隙可以作为未来区域内的公共景观空间，而不同颜色的屋顶通过组合形成一幅生动的抽象绘画。

户型设计

户型设计功能分区明确、合理，最大限度的满足户型的方正和通透性。除客厅、餐厅、厨房、卫生间、卧室以外，着重考虑入口过渡空间、步入式更衣间等辅助空间的合理布置，合理安排各空间的顺序，减少交通面积，提高使用效率。严格组织好公共与私密空间的关系，避免相邻住户的干扰。

设计重点考虑空间的舒适体验，采用餐厅上空设天窗的方式，形成区别于普通住宅的空间开敞感。同时引入双主卧规格等人性化设计元素，为使用者提供尽可能舒适的生活体验。充分考虑各细节的尺寸，符合人体行为学，家具、电器、活动空间、休息空间、娱乐空间等各尺寸张弛有度，恰到好处，提高建筑品质。

五联立面图 -1
Elevation-1 (five-family townhouses)

五联立面图 -2
Elevation-2 (five-family townhouses)

五联立面图 -3
Elevation-3 (five-family townhouses)

五联立面图 -4
Elevation-4 (five-family townhouses)

FEATURE | 专题

双联立面图-1
Elevation-1 (double-family townhouses)

双联立面图-2
Elevation-2 (double-family townhouses)

双联立面图-3
Elevation-3 (double-family townhouses)

双联立面图-4
Elevation-4 (double-family townhouses)

Profile

This project is located in the northwest side of central Xianghe County; the eastern site is close to the main road Daxiang Road of this city, Puchi River flows through the southern site. With a total area of 225,308 m², and its main buildings are townhouse and Larry villa.

Planning and Layout

The project inherited imitating nature, with the design concept of integrating with nature, regarding the earth texture as the planning and design framework by researching the surrounding environment of the site, which developed into abundant planning texture, meanwhile, the layout of the residence combined land using made the spare corner land became public landscape green space, provided communication and recreation place for those resident, which improved the living quality of the whole community.

Architectural Design

Take Holland modern architectural style as the example, the design considered grey body as the base, white balcony as horizontal ties; aside different kinds of materials and 45 degree slope roof hut of different colors on hoar fastens mass, which constitute the elegant and refreshing keynote of the villa community. Intermixed using nature plain red ceramic tile, wood tile, and vertical linear texture ceramic plate, facing brick and other external wall materials, which brought some vibrancy and surprise into the solemn and elegant ashen keynote, and decomposed the three storey building, made it fall to

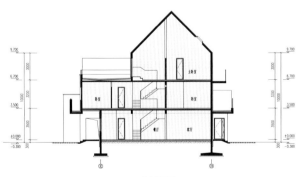

五联剖面图 1-1
Section 1-1 (five-family townhouses)

五联剖面图 2-2
Section 2-2 (five-family townhouses)

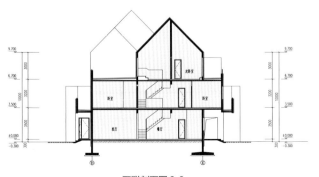

五联剖面图 3-3
Section 3-3 (five-family townhouses)

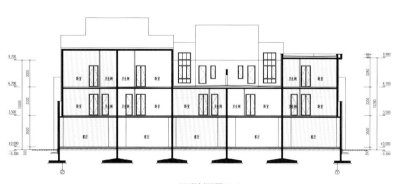

五联剖面图 4-4
Section 4-4 (five-family townhouses)

pleasant scale. The constitution of balcony rail surface, the paving change of gray tiles, the dark coping of white imitation stone paint the wall, upright overlock system of aluminum plate sealing, glass window and other detail design, make the whole building concise and delicate, embodies the pursuit of low and high quality design goal.

Landscape Design

This project created important landscape axis and node, and at the same time also became public life space and important communication platform among neighborhoods through dislocation and facade color of the building, material change, arrangement form and exciting, changeable public landscape space and penetrated in the street. Imitate Piet Cornelies Mondrian 's abstract geometry, permutated and combine the standardized design of residential prototype, the interspace between residence can be used as a future public landscape space, and different colors of roof formed a vivid abstract painting via combination.

House Layout Design

The function of housing type is clear and reasonable, maximally satisfied the permeability of the house. In addition to sitting room, dining-room, kitchen, bathroom, bedroom beyond, mainly considering the transition space of the entry, the reasonable layout of walk-in dressing room and auxiliary space, reasonable arrangement of the space order, reduced traffic area and improved the efficiency. Strictly organize the public and private space to avoid the interference of neighbor residents.

The design considered comfortable experience of the space most, set a skylight up the restaurant, build open feeling that is different from ordinary residential space. At the same time brought in double main bedroom and humanized design elements, provided the user with a comfortable life experience as much as possible. Fully considering the size of the details which is correspond to human body behavior, furniture, electrical appliances, activity space, rest space, entertainment space etc, all the spaces are proper and with loose and tight management , improved the quality of the building.

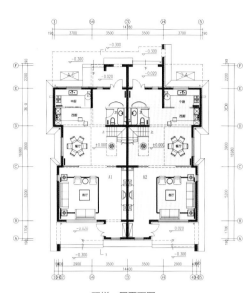

双拼一层平面图
First Floor Plan (double-family townhouses)

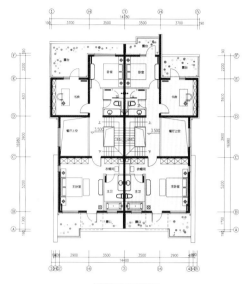

双拼二层平面图
Second Floor Plan (double-family townhouses)

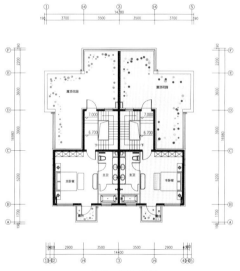

双拼三层平面图
Third Floor Plan (double-family townhouses)

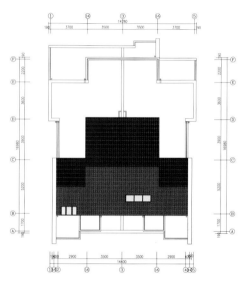

双拼屋顶平面图
Plan for Roof Floor (double-family townhouses)

NEW HOUSE _063

NEW ASIAN-INSPIRED LEISURE COMMUNITY | Nanchang Tianmu • Junhu

新亚洲风情休闲社区 —— 南昌天沐•君湖

项目地点：中国江西省南昌市
建筑设计：筑博设计（集团）股份有限公司
用地面积：163 881 m²
建筑面积：196 616 m²

Location: Nanchang, Jiangxi, China
Architectural Design: ZHUBO DESIGN GROUP CO., LTD
Land Area: 163,881 m²
Floor Area: 196,616 m²

项目概况

项目位于南昌市象湖附近，规划设计以服务城市为设计出发点，生态度假风情文化为主题，规划出一个新亚洲风情主题的休闲社区。

规划布局

项目中酒店定位为生态休闲度假酒店，对城市呈现出开放和包容的态度，结合院落式空间布局特征，将酒店水景、雄溪河景观以及生态公园、庭院绿化等与建筑本身融为一体。住宅设计注重城市生态休闲气质，强调水生态主题，打造湖、溪、湾、桥等水生态景观点，结合建筑布局，体现居住区的水生态健康主题。

建筑设计

建筑风格设计上体现新亚洲休闲地标和多元化需求，希望有相应宽容写意的生活模式可选，体现多元化主题，并满足项目分期开发的需要。

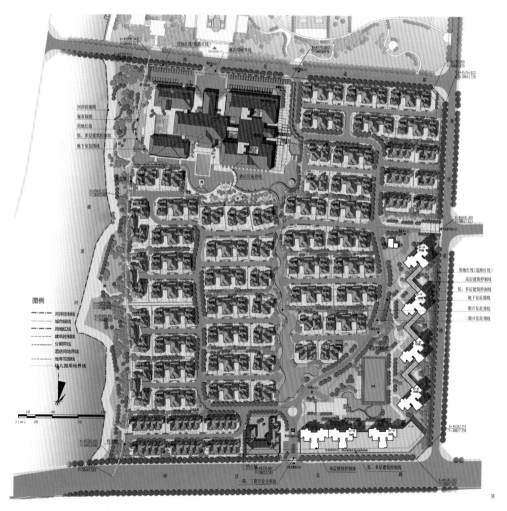

总平面图 Site Plan

Profile

The complex is located near by Xiang Lake in Nanchang city, Jiangxi Province. And serving the city is the starting point for the planning design of this project, the overall theme of which is an ecological area with the culture of vacation feeling, so as to plan out a leisure community on a theme of new Asia customs.

Planning and Layout

The hotel has positioned itself as an ecological leisure resort hotel, which is open and inclusive to this city. Buildings itself, combining the spatial design of courtyard style, integrates with water features in hotel, Xiongxi River landscape, an ecological park and garden greening etc. What is more, the residential design philosophy focuses on the city ecological leisure feeling and emphasizes the theme of water ecology;

负一层平面图
Plan for Basement Floor

一层平面图
First Floor Plan

二层平面图
Second Floor Plan

三层平面图
Third Floor Plan

地下一层平面图
Plan for Basement Floor

一层平面图
First Floor Plan

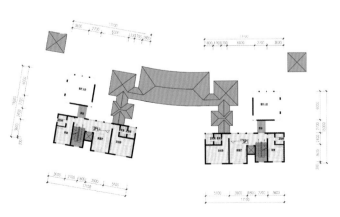

二层平面图
Second Floor Plan

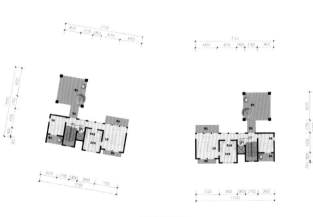

三层平面图
Third Floor Plan

consequently the theme of healthy water ecology is reflected in residential area while incorporating with the architectural layout.

Architectural Design
As for the design of architectural style, it also represents the leisure landmark for new Asia and the diversified demands, and is expected to provide related tolerant and enjoyable life style in order to reflect diversity theme and meet the needs of phased development of the practice for this project.

FEATURE | 专题

AUSPICIOUSNESS, FASHION AND DIGNITY | Purple Garden, Nanjing

紫气东来 时尚尊贵—— 南京（马群）紫园

项目地点：中国江苏省南京市
建筑设计：汉森伯盛国际设计集团
用地面积：110 000 m²
总建筑面积：230 000 m²
容积率：1.12

Location: Nanjing, Jiangsu, China
Architectural Design: Shing & Partners International Design Group
Land Area: 110,000 m²
Gross Floor Area: 230,000 m²
Plot Ratio: 1.12

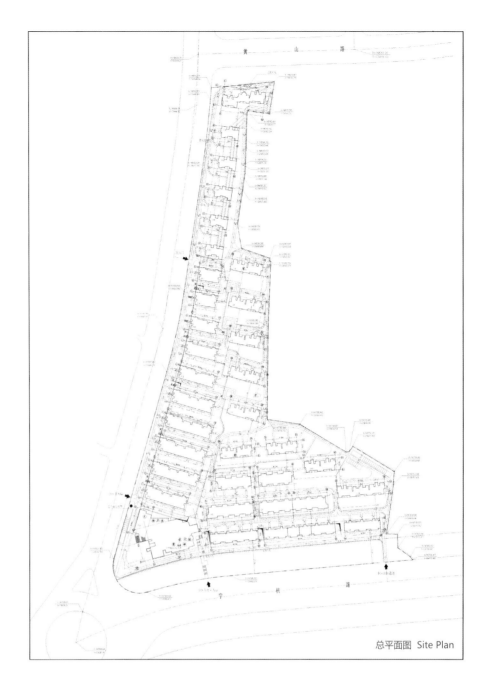

总平面图 Site Plan

NEW HOUSE _073

FEATURE | 专题

轴立面图 Side Elevation

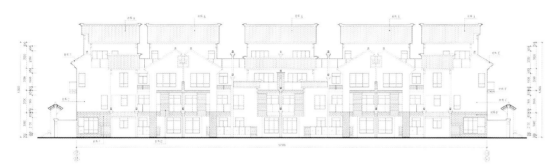

轴立面图 Side Elevation

项目概况

项目地处南京市栖霞区紫京山之东南山麓下,尊南朝北,紫气东来,尽显帝皇气象。

建筑设计

紫园的设计理念早已超越了建造一个居所,而是建造一座满足人对自然、艺术、时尚、尊贵等多重需求的完美居住空间环境。设计中不是死板地照搬传统古建筑中的元素和符号,而是经过精选与提炼,最后探求一种富有深厚东方新人居之韵味,设计出完全符合现代人居住需求的人居环境。从精致到细节,带给住户一个集娱乐、休闲与品位于一体的生活居所、休闲的港湾!

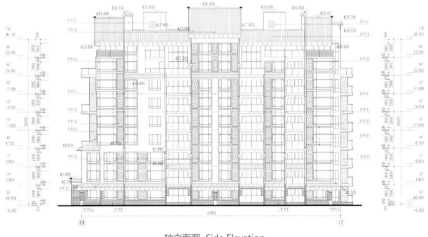

轴立面图 Side Elevation

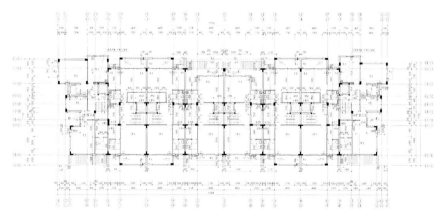

三层平面图 Third Floor Plan

Profile

The project is located at the foot of southeast Zijing Mountain, Qixia District, Nanjing. It is facing north with auspiciousness coming from east, manifesting the imposing manner of an emperor.

Architectural Design

The design idea of the project has been more than building a residence, but building a perfect living space and environment with multiple requirements for nature, art, fashion and dignity. Instead of copying the elements and signs in traditional ancient architectures, the design pursues the new living atmosphere for Orientals through selection and extraction, which leads to the inhabitant environment perfectly meeting the modern living requirement. It has offered the residents an exquisite and detailed destination for living and relaxation, featuring recreation, leisure and taste.

INTEGRATE THE ELEMENTS OF "MOUNTAIN, LAKE, SEA" WITH CONCISE MODERN RESIDENTIAL SPACE

| The First Phase of Shidai Nansha Shanhuhai

融合"山、湖、海"元素的简约现代居住空间

—— 时代南沙山湖海一期

项目地点：中国广东省广州市
建筑设计：深圳华森建筑与工程设计顾问有限公司
总建筑面积：177 700 m²

Location: Guangzhou, Guangdong, China
Architectural Design: Shenzhen Huasen Architectural &
Engineering Design Consultant Ltd
Total Floor Area: 177, 700 m²

项目概况

项目位于广州市南沙街虎门石矿场，距离虎门大桥、南沙客运港仅5分钟路程。基地一面群山环绕、绿树葱葱，一面紧邻入海口狮子洋、碧波荡漾，基地内还有8万m²的天然湖泊，水天一色，空气清新，环境优美。时代南沙项目工程规划总用地面积约36万m²，一期总建筑面积为17.77万m²。

规划布局

总平面依据原有地形、地势，按阶梯状布置住宅，同时沿着海景、中心水景构成的景观带布置建筑物，成为狮子洋边一道独特的风景线。整个住区建筑布局以中心景观湖面为中心顺应地形特质展开，充分利用山体水景景观，并通过建筑高低的排布，使优美的自然形态更加突出地表现出来。这种以自然水系、人工建筑和环境相结合的布局，体现了建筑与空间、空间与环境的和谐，是人与自然、传统与现代的有机结合。

车行交通从东侧市政道路引入本期，沿着小区外侧在区内形成环道。小区的主入口设置在东北侧，正对大面积的中心湖景，也是小区的景观入口所在。小区内人车分流，整个小区采用无障碍通行的设计。道路设计时特别关注了消防车的转弯半径、消防车道及消防扑救面的坡度要求等细节问题。

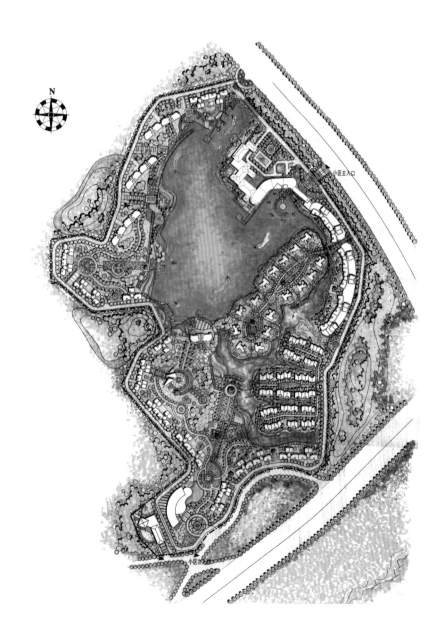

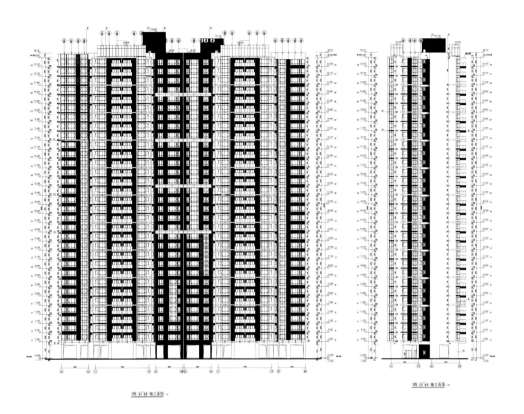

建筑设计

立面设计延续现代而又简洁的风格。将建筑作为环境的背景，主体大面积采用清新的浅色，使之更好的融合于环境中，部分墙面采用棕色、蓝色和绿色，一方面活泼了立面，增加凹凸及层次感，也给建筑赋予了令人舒适的暖色调。空调百叶不仅起到遮挡空调室外机的作用，也是建筑立面的重要构成元素，与凸窗结合在一起，构成竖向元素，与屋顶的框架延伸为一体，加强了塔楼的挺拔感。大面积的转角凸窗和透空的阳台玻璃栏板强调出了建筑的现代气息。

景观设计

项目融山景、水景、海景三景于一体，景观条件非常优越。中心湖湖岸长堤形成的步行景观长廊和丰富的山地景观以及无限广阔的狮子洋海景成为最富特色的自然景观带，充分表达尊贵优雅的居住环境与自然、生态、健康、环保的设计理念。

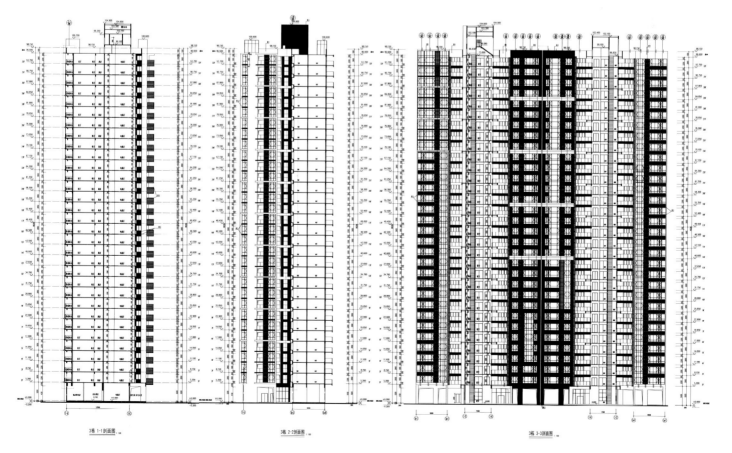

3栋 1-1剖面图　　　3栋 2-2剖面图　　　3栋 3-3剖面图

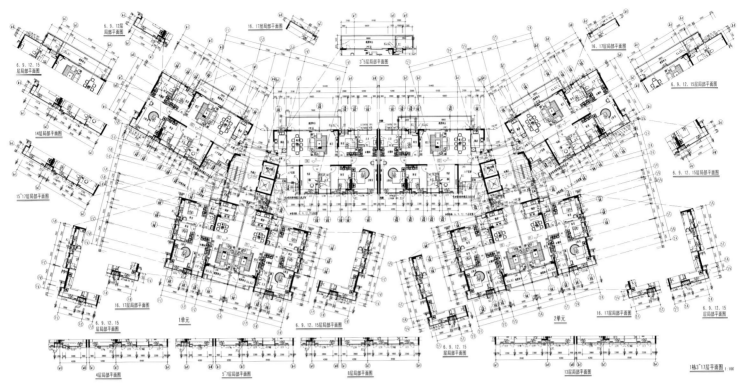

标准层平面图 Plan for Standard Floor

Profile

The project is located in Humen Quarry on Nansha Street, five minutes drive from Humen Grand Bridge and Nansha Passenger Port. The site faces mountains and lush green trees on one side and closely next to Lion Ocean Sea Mouth on another side. Inside the site, there is one 8,000 m² large natural lake, where people could enjoy the same color of water and sky, fresh air and fine environment. The overall planning area of Modern Nansha is 360,000 m² while the first phase takes up 177,700 m².

Planning and Layout

The general layout of residence spreads in a terraced form according to the original terrain conditions while arranges buildings alongside the landscape belt of sea and central lake landscape, which altogether become a special landscape line of Lion Ocean. With central lake as the core of landscape and the surrounding terraced terrain, the project highlights the original fine topography through taking advantage of mountains, waters and different levels of architectures. This layout which integrates natural water systems, manmade buildings and environment demonstrates the harmony of the three major elements and the organic combination of people and nature, tradition and modernity in a broader sense.

Vehicle stream is led into the side from municipal roads on the east side, which form a ring shape outside the community. The main entrance is arranged in the northeast corner, facing directly towards the central lake–the landscape entrance of the community. Inside the site, pedestrians and vehicle roads are separated for safety reason and barrier-free design is adopted at the same time. Other detailed problems were also taken into consideration, such as the turning radius of fire truck, fire truck lane and the slope requirement of firefighting surface.

Architectural Design

The facade design continues modern and simple style, applying large area of fresh and light color to the main body and using buildings as the backdrop of environment, which blends the buildings into the surrounding environment. Part of the walls adopts brown, blue and green, which activates the elevation and adds sense of concave-convex and gradation on one side and puts on a warm tone to the buildings. The louvers not only block the outdoor unit of air-conditioners, but also perform as an important element of architectural elevation, joining the projecting windows to become vertical elements. They also stretch as one with the rooftop frame to increase sense of tall and straight. Large area of corner projecting windows and hollow terrace glass boards stress the modern temperament of the buildings.

Landscape Design

The project embraces mountain, water and sea landscape all in one site, offering superior landscape to residents. Central lakeshore bank promenade corridor, the abundant mountainous landscape and unlimited Lion Ocean landscape become the featured natural landscape belt, fully expressing the design concept of elegant, dignified residential environment-natural, ecological, healthy and environment friendly.

ANCIENT DIVERSIFIED LANDSCAPE SPACE COMBINING CHINESE AND WESTERN ELEMENTS

| Foshan Hefeng Yingyuan Garden

中西合璧、古韵有致的多元化景观空间 —— 佛山和丰颖苑

项目地点：中国广东省佛山市
开 发 商：佛山和丰颖苑房地产有限公司
景观设计：广州邦景园林绿化设计有限公司
占地面积：65 464 m²
建筑面积：118 111 m²

Location: Foshan, Guangdong, China
Developer: Foshan Hefeng Yingyuan Garden Real Estate Co., Ltd.
Landscape Design: Guangzhou Bonjing Landscape Design Co., Ltd.
Land Area: 65,464 m²
Floor Area: 118,111 m²

项目概况

和丰颖苑位于佛山市南海区丹灶镇仙湖旅游度假区的仙湖之畔,占地面积约65 464 m²,建筑面积约118 111 m²,整体绿化率为35%,容积率为1.2。规划的住宅小区总户数约900户。户型面积区间由36 m²的一房至180 m²的四房组成,更有少量约260 m²复式单位。

规划布局

项目共分为三个地块,东面为仙湖酒店、无极养生园公园,南面为桂丹路,西面为别墅小区,北面为仙湖旅游度假区及仙湖湾广场。小区采用南北朝向,根据地块形态半围合布局及排列式,以南北向户型为主,部分东西向。小区内沿中间地块设南北轴线,另一条轴线沿着桂丹路平行通过水系及溪流景观延伸,使三块地块有机联系形成一个整体。

景观设计

整体设计采用了"中西合璧,古韵新做"的基调处理,借鉴了中国古典园林的空间营造手法,并以西方现代的景观语言加以诠释。以自然生态原则为主导,尽可能增加绿化面积和个性多样化的景观空间,使得建筑群体与花草林木共生共融,为业主们提供丰富的景观视野和放松身心的社区空间。

同时运用了步移景异、曲径通幽、植物围合等造园手法,营造出丰富的景观序列空间,通过多重的地形塑造层层递进的空间感受,营造出围合的或开敞的、公共的或私密的、凸起的或下沉的、通透的或隐蔽的多样空间,以静止的空间组合场景,以流动的时空牵动游人的步伐,让景物依次展开。利用植物、水景、铺装、小品等营造空间,并通过有趣味性的路径相连,为居民提供休息、活动、健身等不同功能的活动场所,力求达到"庭院深深深几许"的意境,方寸间自成天地,精细自然之处却包含无限深邃。

社区景观布局精妙,层次高低错落,道路回环,空处有景,疏处不虚,大中阔景,小中致景,密而不逼,静中有趣,幽而有芳。绿化种植形成绿树成荫、繁花似锦、曲水回环、花堤柳岸等具有特色的景观韵味。入口处,以极富动态的水景设计诠释了古典的奢华魅力,以其灵动的跌水为宁静的社区迎来汩汩生机。园内设计了景观湖,湖岸细柳轻摇、暗香浮动、湖面波光粼粼、疏影横斜,湖中喷泉直涌,结合湖边廊、亭、栏,巧妙地构成一体。整个景观设计做到了"布局之工,结构之巧,装饰之美,营造之精,文化内涵之深"的境界。

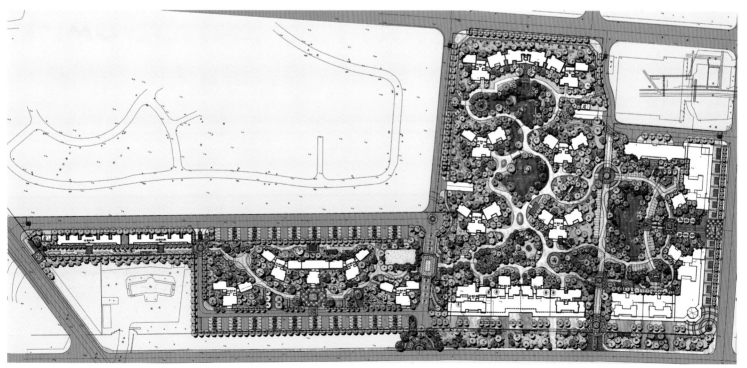

总平面图 Site Plan

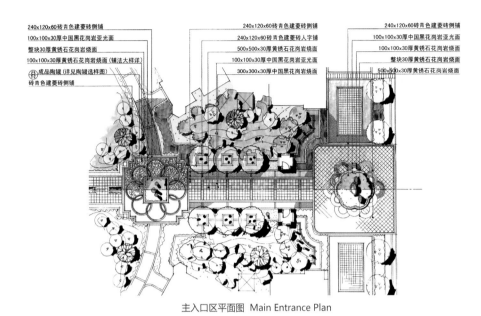

主入口区平面图 Main Entrance Plan

Profile

Hefeng Yingyuan Garden is located near the Xianhu Lake of Xianhu Tourism Resort District, Nanhai district, Foshan city. It covers an area of about 65 m², building area of about 118 m² and the overall greening rate is 35%, plot rate is 1.2. There are about 900 households. These houses vary from 36 m² a room to 180 m² four-room suite, and a small amount of 260 m² multiple units.

Planning Layout

The project is divided into three blocks. There are Xianhu Hotel, Wuji Well-being Park in the east, Guidan Road in the south, villa communication in the west, Xianhu Tourist Resort and Xianhu Bay Plaza in the north. This residential garden sits north and faces south. According to landform, the garden adopts half enclosing layout and arrangement type, giving priority to north-south

apartment layout and a few east-west households as well. The garden has a north-south axis in the middle block and the other axis along Guidan Road, paralleling through the drainage net and stream landscape to connect these three blocks into an organic entirety.

Landscape Design

The overall design adopts a key note of "combining the Chinese and Western elements, taking new ideas from the past", using the reference of garden construction technique in Chinese classical landscape architecture and impressing it with the western modern landscape language. Led by the natural ecological principle, the green area and personality-diverse landscape space are added as much as possible to make the building group and plants symbiotic and harmony and provide the owners abundant landscape view and physically and mentally relaxed space.

At the same time, it uses the landscape techniques such as changing views in every step, winding path leading to a secluded spot and plant enclosing to build a rich landscape sequence space. Through multilayer terrain of progressive sense of space, a closed or open, public or private, raised or sinking, connect fully or concealed space is built. The stillness of the space combines the scene and the flow of time and space affect visitors' pace. Besides, it uses plants, water, pavement, sketch, etc to create space, and through the funny linked path to provide the resident an activity place for rest, activities, fitness and so, striving to achieve the artistic conception of garden.

The landscape layout is subtle with level-heights at random and winding roads. It sets views in details, tight and not forced, quiet and interesting. The green trees make a pleasant shade and flowers blooms like a piece of brocade, water runs in circle, the flower and willow flourish along the shore. At the entrance, the highly dynamic waterscape design interprets the classic luxury charm. Every active drop of water adds the quiet community more vitality. There is also a landscape lake, fine willow jiggling at the shore with floating fragrance. The lake sparkles in the sunlight and the fountain in the middle of the lake rushes up straightly, ably blending with gallery, pavilion and hurdles near the lake. The whole landscape design has reached the state of "neatly layout, artful structure, decorative beauty, refine design and deep cultural connotation ".

FEATURE | 专题

MOUNTAINS AND WATERS SYMPHONY OF AUSTRIAN EXPRESSION

| Wukuang Hallstatt Villas

奥地利风情演绎的山水交响曲—— 五矿•哈施塔特别墅区

项目地点：中国广东省惠州市
开 发 商：博罗碧华房地产开发有限公司
规划/建筑设计：美国德空建筑设计有限公司
景观设计：英国阿特金斯集团
主创设计：GALE JOSEPH REIBEL、刘一玮、王晖、邹嘉盟
总用地面积：1 000 000 m²
总建筑面积：1 000 000 m²
容 积 率：1.0
绿 化 率：45%

Location: Huizhou, Guangdong, China
Developer: Boluo Bihua Real Estate Development Co., Ltd.
Planning/Architectural Design: TECON LCC
Landscape Design: ATKINS
Chief Designer: Gale Joseph Reibel, Liu Yiwei, Wang Hui, Zou Jiameng
Total Land Area: 1,000,000 m²
Total Floor Area: 1,000,000 m²
Plot Ratio: 1.0
Greening Ratio: 45%

项目概况

项目位于惠州江北CBD都市生态新区，广东省惠州市博罗县城、汤泉度假区内。东距惠州市政府约12km，西距博罗县城5km。本地块处于罗浮山山脉，三面环山，170 000m² 天然活水湖泊环抱其中，占总用地的约20%，风水上呈"富阴抱阳"态势。总规划用地约1 000 000m²，共分6期开发，目前正在开发第一期。第一期占地360 000m²，产品类型丰富，规划有双拼别墅、联排别墅及奥地利风情小镇形态的商业休闲中心。

规划布局

以中心湖和奥地利风情小镇为核心，环绕内外两层环路，以及由湖岸放射延伸到山地的几条景观轴线，支撑起整个住区的空间骨架。

道路系统层级分明，考量整体营造、管理及景观效益，将道路系统分为主干道、汇集道路

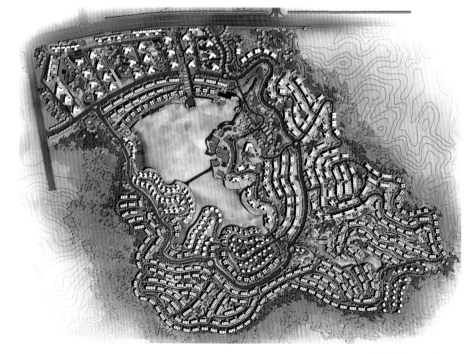

总平面图 Site Plan

NEW CHARACTERISTICS | 新特色

轴立面图 Side Elevation

轴立面图 Side Elevation

及小区入户道路。营造优质主干道景观意象，做为宅基地进出道路；以循着河谷而行的汇集道路串连各小区，营造清楚的动线；各小区则以独立入户道路，塑造完整的小区意象与自明性。

住宅建筑依据山形地势铺展，具有良好的景观视线。几乎每个单体建筑都有天然的高差，从而赋予每个单体独特的个性和归属感。

建筑景观一体化设计

以音乐为设计主题，把奥地利独特而高超的文化艺术内涵和建筑元素融合在建筑和环境空间的设计里，以期能为在这里居住和游览的人们带来丰富而深刻的生活体验。华尔兹大道、音乐广场、海顿码头等，这些以音乐为主题的景观空间让音乐流淌在园区的各个角落。

建筑选用极具独特性的奥地利城镇风格，并把这一风格贯穿在空间组合、尺度把握和细部造型中。从手法到材料运用的精准选择，把奥地利乡村朴素优雅、和谐欢乐的特质作为艺术表现的内涵。遵从城镇建筑的宜人尺度，加上独具个性的抹角四坡屋顶、精致的门窗套、精美的木质花饰檐口与栏杆，这些体现精美工艺水准的细部完全不同于司空见惯的豪华欧式风格，给人耳目一新的感受。

⑤ ④ ①

轴立面图 Side Elevation

NEW CHARACTERISTICS | 新特色

户型设计

住宅单体设计紧密配合山地规划的思想，把山地住宅的优势淋漓尽致地发挥出来。层层退台的格局使各户拥有无遮挡的宽广视野；双首层入户设计，使得前后庭院巧妙连接。同类户型通过精心调配的外墙色彩和材料及细部精巧造型，加上依借不同的地形地势，呈现出户户不同的专属品质。

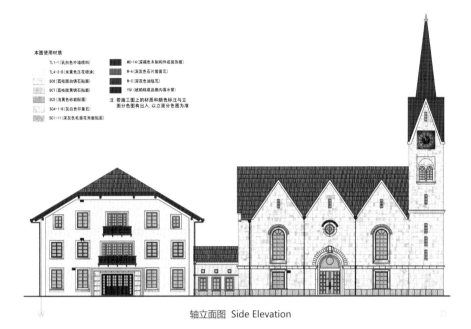

轴立面图 Side Elevation

立面图 Elevation

立面图 Elevation

立面图 Elevation

立面图 Elevation

NEW CHARACTERISTICS | 新特色

Profile

The project is located in the CBD urban ecological area north of Jiangbei of Huizhou, inside the Hot Spring Resort of Boluo County. The site is 12 km west of Huizhou and 5 km east of Boluo County with Luofu Mountains embracing from three sides. A 170,000 m² natural fresh water lake is inlayed in the site, covering 20% of the entire site, which demonstrate a "rich Yin embrace Yang" pattern according to Fengshui. The project is divided into six phases and Phase One is under construction now, which covers an area of 360,000 m² with abundant buildings types, from two-family house, townhouse to commercial and leisure center of Austrian style.

Planning Layout

With central lake and Austrian town as the center, the entire community spatial skeleton is supported by two circling roads and a few landscape axis radiating from the lake bank to mountains.

The clearly terraced circulation system takes into consideration of overall magagement and landscape effectiveness, consisting of main stems, collection roads and community indoor paths. The main stem network is not used for entering or exiting residence; the collection road network alongside the valley strings each community and forges clear motion generatrix; each community enjoys individual and independent indoor path, which helps create a complete community image and self-evidence.

The residence stretches along the mountain terrain, providing fine landscape visual lines. Almost every single building possesses natural height difference, thus offer them each unique sense of identity and belonging.

The Integrative Design of Building and Landscape

With music as the design theme, the project integrates unique and superb cultural and artistic connotation with buildings and the environment to provide rich and deep life experience for people living here. Landscape spaces such as Waltz Avenue, music square and Haydn wharf etc. all flow with music in every corner.

The project adopts special Austrian town style and penetrates it into spatial combination, grasping scale and detail modeling. From the exquisitely selection of techniques and materials, the elegance and harmony traits of

Austrian country elements are taken as artistic connotation. The project abides by the pleasant scale of urban buildings and adds distinctive hipped roof with wiped angle, delicate door and window cover, elegant wood Floriation cornice and railing, all of which differ from the common luxury European style and provide a completely fresh feeling.

House Layout Design

Single room type of the project is closely coordinated with the idea of mountainous region planning and fully gives play to the advantage of mountainous residence. The terraced stepping back pattern provides each family with unlimited broad view. The double first floor design ingeniously connect the front and back courtyards together. Same room types demonstrate exclusive feature with the carefully blended color on the exterior elevation and materials, exquisite modeling in details and different terrains.

SIMPLE AND COMFORTABLE, EXQUISITE AND ELEGANT

| B1-A Model Apartment of ZhiyuLanwan Garden in Shenzhen

简洁舒适 精致优雅 —— 深圳联路芷峪澜湾花园B1-A户型样板房

项目地点：中国广东省深圳市
客户：深圳市联路投资管理有限公司
室内设计：戴勇室内设计师事务所
软装工程/艺术品：戴勇室内设计师事务所&深圳市卡萨艺术品有限公司
建筑面积：90 m²
摄影师：江国增
使用物料：希拉克木纹云石、银灰洞云石、透光云石、灰影木饰面、墙纸、清镜
设计时间：2011年
完工时间：2012年7月

Location: Shenzhen, Guangdong, China
Client: Shenzhen Lianlu Investment Co., Ltd.
Interior Design: Eric Tai Design Co.,LTD
Soft Engineering/Artifacts: Eric Tai Design Co., Ltd. & Shenzhen Kasa Artifacts Co., Ltd.
Floor Area: 90 m²
Photography: Jiang Guozeng
Materials: Jacques chirac marble grain, aluminium hole marble, light translucidus marble, gray shadow wood veneer, wall paper, clear mirror
Design: 2011
Completion: July, 2012

浅色的灰影木饰面围合出清新雅致的居家氛围，虽是才90 m²的两层小复式，却一样可以拥有如此丰富的内容和精致的细节，折射出生活的方方面面。

镜面与灰影木饰面的搭配，简约中散发出清新时尚的格调。透着暖黄色灯光的云石台面，沙发背景墙从顶上垂下来的镜面不锈钢条，让人眼前一亮。精致优雅的饰品和摆设，哪怕是一盆花，一把汤匙，一只酒杯，都令人爱不释手，让人不由自主的感受到主人的生活情趣。主人房的背景墙，通过软包的分缝，再加一盏水晶吊灯和一幅精选的挂画，便成就一组和谐的构图，简洁却舒适。

整套样板房给人的感觉就像一位淡妆的少妇，虽没有浓妆艳抹的魅惑，却亲切而舒适，散发着清淡优雅的品味。

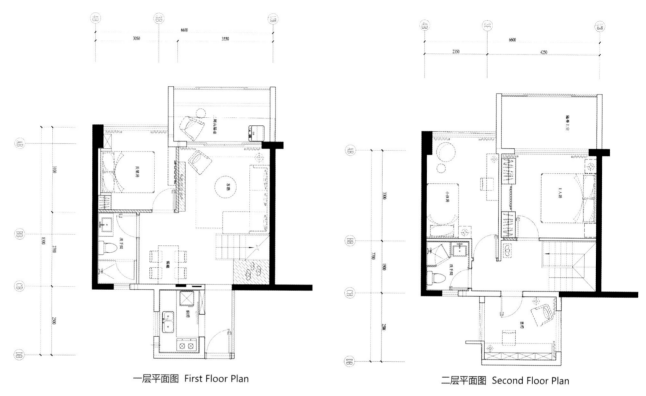

一层平面图 First Floor Plan   二层平面图 Second Floor Plan

The light grey shadow wood veneer creates fresh and tasteful residential atmosphere. A 90 m² double-storey loft could also uphold such abundant content and exquisite details and reflect the many aspects of life.

The collocation of mirror and grey shadow wood veneer spreads fresh and fashionable style out of simplicity; the marble counter penetrating yellow warm light and stainless steel strips hanging on the sofa background wall light up people's eyes. The exquisite and elegant decorations and furnishings, even a flower, a spoon or a wineglass all make people fondle admiringly and think of the host's spice of life. The background wall in host's bedroom is wrapped with soft parting, with a crystal chandelier and a carefully selected picture, forming a harmony, simple and comfortable picture.

Without the enchantment of heavy powdering, the complete model apartment reminds of a young woman with simple make-up, amiable and comfortable, expressing the taste of delicacy and elegance.

MULTIPLE HEIGHT, RHYTHMICAL RIVERFRONT HOUSE

| Riverfront Housing, Santa Coloma de Gramenet (Phase II)

多重高度 极富韵律感滨江住区

—— 圣·科洛马·德·格拉曼历镇滨江住宅项目二期

项目地点：西班牙巴塞罗那圣•科洛马•德•格拉曼历镇	Location: Santa Coloma de Gramenet, Spain
开 发 商：圣•科洛马•德•格拉曼历镇政府	Developer: Town Hall of Santa Coloma de Gramenet
建筑设计：Ravetllat Ribas	Architectural Design: Ravetllat Ribas Arquitectura
合作设计：Olga Schmid建筑设计事务所 STATIC结构工程 Briz Fumadó, M&E工程 Ángel Rodríguez, José Luís de la Fuente, Site surveyors	Collaborators: Olga Schmid, Architects/ STATIC Ingeniería, Structural Engineers/ Briz Fumadó, M&E Engineers/ Ángel Rodríguez, José Luís de la Fuente, Site surveyors
面　　积：27 545.16 m²	Area: 27,545.16 m²
摄　　影：Adrià Goula	Photographer: Adrià Goula

项目概况

该建筑为三大住宅楼群二期项目，位于建筑群西面也是三栋建筑中离江最近的一栋。建筑被四个阶梯分隔成四个单元。设计旨在通过清晰的规划以及不同高度形成双重空间，同时又加强建筑与贝斯河的联系。

建筑设计

建筑高度达17.6 m并分为三大结构区。中心部分为通风露台区并作为交流服务区，立面在连接内外起着过滤器的作用并有效的节约了空间。外立面有双重含义，彩色的面板使建筑增添了不少韵律，另外一个含义在于砖墙，与南面形成斜坡并使整体有一种方向感。

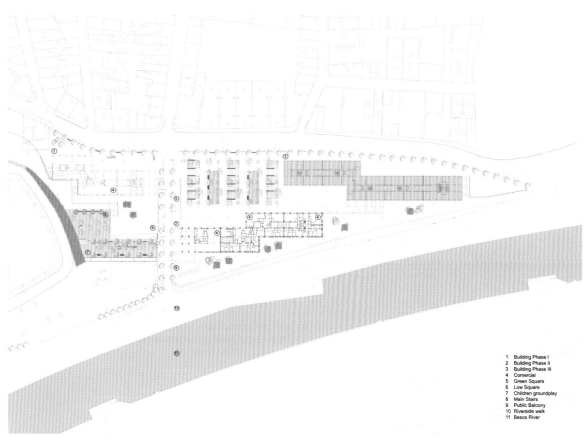

1. Building Phase I
2. Building Phase II
3. Building Phase III
4. Comercial
5. Green Square
6. Low Square
7. Children groundplay
8. Main Stairs
9. Public Balcony
10. Riverside walk
11. Besos River

总平面图 Site Plan

NEW IDEA | 新创意

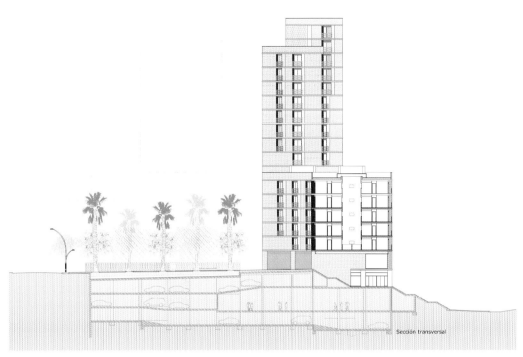

剖面图 Sectional Drawing

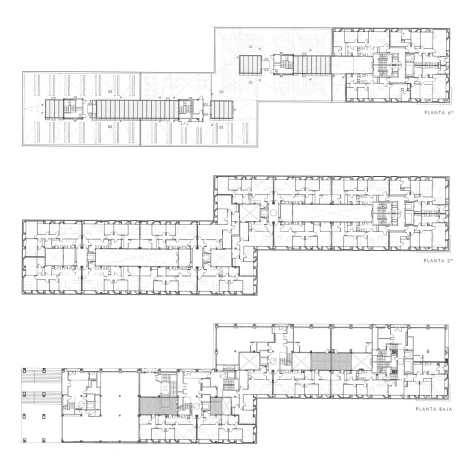

PLANTA 6ª

PLANTA 3ª

PLANTA BAJA

NEW IDEA | 新创意

Profile

The building is the second in a project of three housing blocks, this second building is situated to the West and is the nearest building to the river of all three. It is organized into four staircases with four flats per landing. With the articulation of the plan and its different heights, the building tries to assume a double condition, simultaneously establishing relations with the Besòs River and the small existing buildings.

Architectural Design

The building has been designed with a relatively large depth of 17.60 meters and is organized in three structural bays. The central one is reserved to services and communication hubs which are orientated to covered and ventilated patios. The facades set a filter between the interior and exterior and provide efficient storage space. The materiality of the facade has a double meaning. One is defined with a coloured panel that gives the building a compositional rhythm. The other is the brick wall, which always slopes to the South and gives a directional sense to the whole.

Vivienda tipo. Esquina

Vivienda tipo. Giro

Vivienda tipo. Barra

COMME BUILDINGS

RCIAL 商业地产

P112
国家开发银行三亚研究院：
临海环山 功能多元的综合性建筑

P116
深圳星河龙岗COCO PARK：
高品质 一站式购物中心

P128
深圳创维石岩科技园二期：
都市型工业综合体

公司简介

华艺设计顾问有限公司1986年在香港注册,同年在深圳独资设立香港华艺设计顾问(深圳)有限公司,设有上海、南京、武汉、北京、重庆、广州、厦门和成都等分公司,是具有甲级工程设计资质的工程设计咨询企业(证书号为:AW144017173)。2009年在北京注册成立北京中海华艺城市规划设计有限公司,2010年8月获得国家住房与城乡建设部颁发的"城乡规划编制甲级资质证书"[建]城规编第(101205)。

华艺拥有数百人的高质素专业设计团队,设有规划、建筑、结构、强电、弱电、给排水、暖通空调、总图、概预算、室内设计等专业,主要承接各类公共与民用建筑工程设计、城市设计、居住区规划与住宅设计、室内设计及前期顾问和建筑策划研究等业务。公司凭借先进的管理,国际视野的底蕴,突显出强大的竞争力。发展足迹已遍布全国。

城市综合体,作为提升城市价值及形象的重要载体与标志,已成为当今地产行业发展的主流趋势,引领了集约型城市开发的潮流。每个"城市综合体"都融合了办公、商业、酒店、居住、娱乐、休闲等多种功能,是一个高效率的有机集约系统。华艺公司凭借其出色的综合能力,在全国完成了多个重要的城市综合型建筑的设计工作,如:深圳星河时代广场、中航天逸花园、中海珠海富华里中心、济南中海广场•寰宇城、贵阳未来方舟C区环球谷等10余个超大型城市综合体项目。这些项目通过其自身的丰富业态空间组合,互为依托,互为促进,共同分享彼此价值的提升,形成了一个自成体系的全新价值平台。

COMPANY PROFILE

HUAYI Design Consultants Ltd. (HUAYI) registered in Hong Kong in 1986, and in the same year set up the wholly-owned company in Shenzhen, named Hong Kong HUAYI Design Consultants (Shenzhen) Ltd. HUAYI today is an engineering consultant enterprise with Grade A Construction and Engineering Design License (Certificate ID: AW144017173), with branches in Shanghai, Nanjing, Wuhan, Beijing, Chongqing, Guangzhou,Xiamen and Chengdu. In 2009, China Overseas Huayi Urban Planning & Design Co., Ltd.(Beijing) was set up, and in August, 2010, it was awarded "Grade A Urban-Rural Planning License" by MOHURD.

HUAYI has hundreds of personnel of highly-educated, high-quality and experienced, the design team has various professions of planning, building, structure, heavy-current, light-current, water supply and drainage, HVAC, general layout, engineering budget and interior design, mainly undertaking businesses of public and civil construction design, urban design, residential planning and design, interior design, pre-consultancy and construction strategy researches. HUAYI showed the outstanding competence with advanced organization and international sights.

Urban complex, as an important carrier for city value and the landmark of a city, has already become the hot spot in real estate industry, which leads the trend of integrated urban development. Every urban complex usually integrates the functions sch as office, retail, hotel, residence, entertainment and recreation, forming a high-efficient architectural system. With outstanding competence and strong power, HUAYI has completed the design for about ten important urban complexes all around China, for example, Galaxy Times plaza in Shenzhen, Tianyi Garden, Lopburi Center in Zhuhai, UNI in Jinan, Global Valley of Future Ark in Guiyang, etc. These projects are well designed with different programs depending on each other and serving each other. Thus a new value system is created.

商业综合体代表作品

MULTI-FUNCTIONAL COMPLEX NEARBY MOUNTAINS AND THE SEA

| National Development Bank Sanya Research Institute

临海环山 功能多元的综合性建筑 —— 国家开发银行三亚研究院

项目地点：中国海南省三亚市
建筑设计：香港华艺设计顾问（深圳）有限公司
建筑面积：14 287 m²

Location: Sanya, Hainan, China
Architectural Design: Huayi Design & Consultant (Shenzhen) Co., Ltd.
Floor Area: 14,287 m²

项目概况

国家开发银行三亚研究院是集高标准客房、餐饮娱乐、休闲度假为一体，具备企业研修培训与国际会议功能的综合性建筑。项目用地位于海南省三亚市三美湾，一面临海，三面环山。折线型的动感建筑得到了多角度的海景视野；木格栅的建筑表皮起到了遮阳节能的效果。

总平面图 Site Plan

COMMERCIAL BUILDINGS | 商业地产

Profile
The complex housing high-standard guestrooms, restaurants, entertainment and resort facilities, can be used for research, training and international convention. The site is located nearby Sanmei Bay of Sanya City with the Sea on one side and mountains on the other three sides. Folding and dynamic form provides more sea views, and the wooden grilling skin can well block the sun and save energy.

HIGH-QUALITY, ONE-STOP SHOPPING CENTER | Xinghe Longgang COCO Park

高品质 一站式购物中心 —— 深圳星河龙岗COCO PARK

项目地点：中国广东省深圳市
建筑设计：香港华艺设计顾问（深圳）有限公司
总建筑面积：172 000 m²

Location: Shenzhen, Guangdong, China
Architectural Design: Hong Kong Huayi Design Consultants (Shenzhen) Co., Ltd.
Total Floor Area: 172,000 m²

项目概况

龙岗COCO Park是星河集团投资10亿元打造的深圳东部首席家庭型消费一站式购物中心。项目位于龙岗商业文化中心区深惠路与黄阁南路交汇处，占地面积30 000 m²，总建筑面积172 000 m²，地上四层，地下四层。汇聚国际仓储式超市、巨幕影院、真冰场、时尚服饰、特色美食、精品家居、潮流数码等七重业态，力求打造成一站式集合奢华购物、休闲、娱乐、餐饮、运动、商务、教育、亲子等八大功能于一体的高品质、现代化购物中心。在龙岗还缺乏大型购物中心的时代，COCO Park的崛起，必将成为龙岗商业新中心最为耀眼的商业名片。

COMMERCIAL BUILDINGS | 商业地产

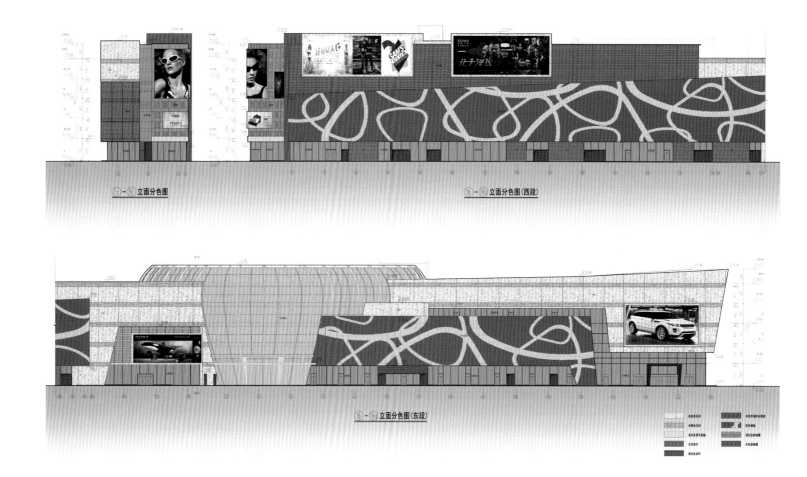

Profile

Longgang COCO Park is the first one-stop family shopping center in east Shenzhen, which is constructed by Xinghe Group with an investment of ten billion. It is located in the intersection of Shenhui Road and South Huangge Road of the Longgang Commercial & Cultural Center, bearing a land area of 30,000 m², and a total floor area of 172,000 m², four floors above ground and four floors underground. Combining the international srockroom-type supermarket, IMAX cinema, skating rink, fashionable costume, specialty, boutique home furnishing and tide digital, it strives to become a one-stop, high-quality and modern shopping center with luxurious shopping, leisure, entertainment, catering, sport, business, education and parent-child activity. In the time of lacking large shopping mall in Longgang, the rise of COCO Park is certainly to be the most shining star of the Longgang New commercial center.

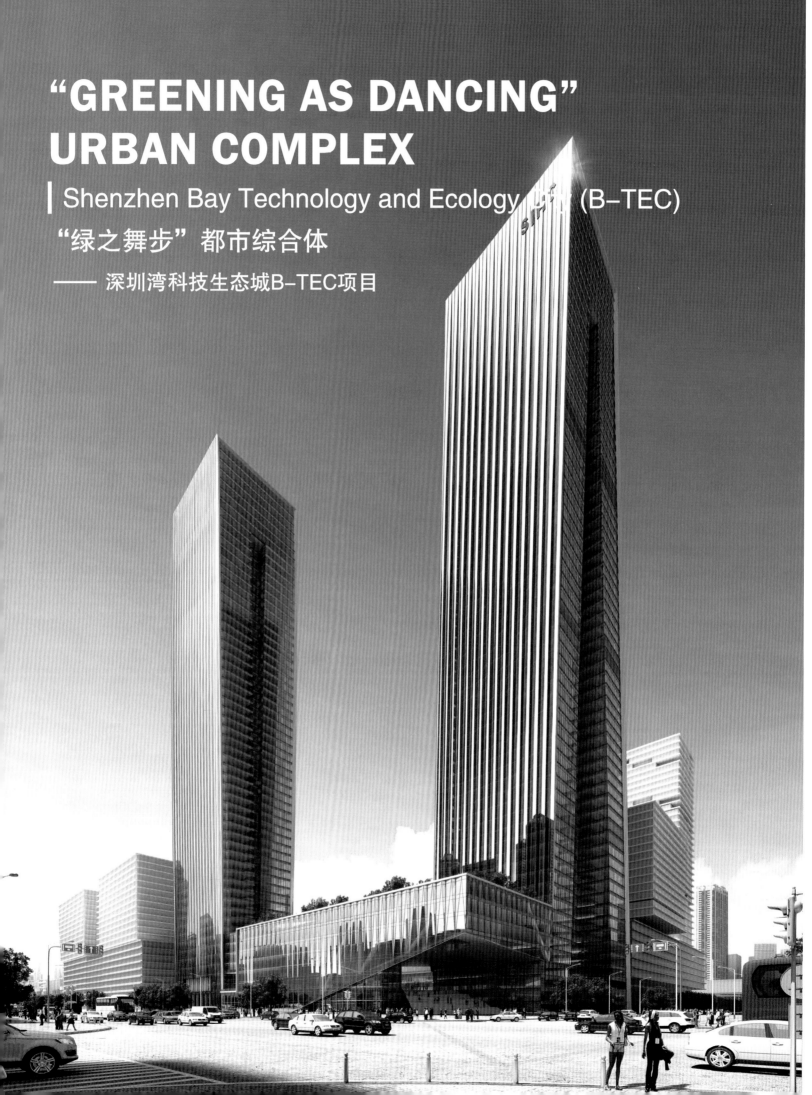

"GREENING AS DANCING" URBAN COMPLEX

Shenzhen Bay Technology and Ecology City (B-TEC)

"绿之舞步"都市综合体
—— 深圳湾科技生态城B-TEC项目

项目地点：广东省深圳市
建筑设计：香港华艺设计顾问（深圳）有限公司
总建筑面积：420 000 m²

Location: Shenzhen, Guangdong
Architectural Design: Hong Kong Huayi Design Consultants (Shenzhen) Co., Ltd.
Total Floor Area: 420,000 m²

项目概况

本项目做为深圳湾科技生态城第四标段超高层项目，位于深圳市南山区高新技术产业园区南区，主要由两栋250 m的超高层塔楼组成。项目总建筑面积约为420 000 m²，是一座由办公、酒店、商业、会议中心复合而成的都市综合体。

建筑设计

方案以"绿之舞步"作为设计原点，通过塔楼至上而下的微妙错动，与裙房形成连续有机的拓扑关系，形如踏歌而来的探戈舞者，奏响了飞扬激昂的城市旋律。

超高层塔楼平面东西错动，利于改善周边区域风环境，同时可获得更多的南北采光面。裙房通过引入生态中庭，巧妙化解大体量建筑通风、采光的不利因素；西北、东南的架空处理，勾勒出建筑群鲜明大气的主入口形象。

Profile

The project is the forth mega-high-rise building in Shenzhen Bay Technology and Ecology City, located in the South Area of Shenzhen Nanshan High Technology Industry Park, consisting of two 250 m mega-high-rise buildings. It occupies a total floor area of 420,000 m², combining office, hotel, business and meeting to be an urban complex.

Architectural Design

The design of the project is based on the concept of "greening as dancing" that means the topological relation of the interaction from top to bottom of the tower and the annexes just like a tango dancer playing an excited city melody.

The interaction from top to bottom of the tower helps to improve the surroundings and obtain more lighting in the north and south orientation. The introduction of annexes in the courtyard subtly defuses the negative factors of ventilation and lighting in such a huge building; the superterranean design in the northwest and southeast draws the distinctive and magnificent image of the main entrance.

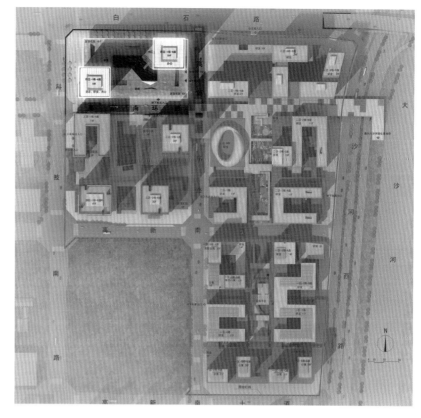

总平面图 Site Plan

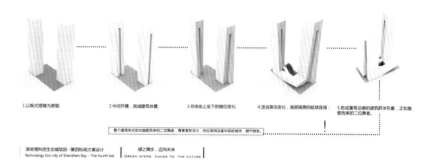

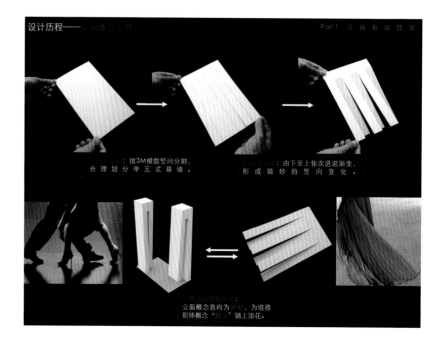

COMMERCIAL BUILDINGS | 商业地产

A COMPLEX WITH VARIATIONS AND CHARACTERISTICS OF TIMES

Shenzhen Media Group Cable TV Hub Building

富于变化 独具时代特征的集合体
—— 深圳广电集团有线电视枢纽大厦

项目地点：广东省深圳市
建筑设计：香港华艺设计顾问（深圳）有限公司
总用地面积：8 661.1 m²

Location: Shenzhen, China
Architectural Design: Hongkong Huayi Designing Consultants (Shenzhen) Co., Ltd.
Total Land Area: 8,661.1 m²

新、老建筑的结合，同样是设计的重要着眼点。我们不仅仅局限于老楼外观的翻新及与新楼功能的简单连接，而是创新性的将新楼的表皮及功能空间与老楼自然的过渡，从而达到功能的统一与形象的完整。

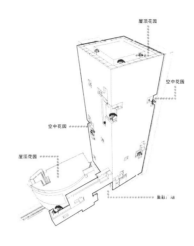

新老建筑结合
New old construction union

项目概况

深圳广电集团电视枢纽大厦（暂定名）项目基地位于深圳市福田区彩田路西侧、莲花支路南侧的B306-0005号地块，距深圳中心区核心直线距离约2 km，属中心区北延的"泛中心"区域。项目占地约8 661.1 m²，设计高度100 m（23层）。项目建成后，主要作为深圳广电集团和天威视讯的办公场所。

规划布局

本项目用地紧张，空间局促，底层架空的方式可有效地缓解场地拥挤所带来的压迫感。阶梯状的景观处

理，消化掉东西的高差，使得空间更为流畅、通透，人们在此交往、停留，成为信息互通的场所，同时它打通了内部空间与城市空间之间的隔阂，加强了莲花山及笔架山两个城市公共景观与使用者之间的联通性。

建筑设计

在该办公楼的设计中，使用空间的舒适性是建筑设计的出发点。交通核的外置以及采用新型的结构形式，创造出一个开敞、无遮挡的办公区域。

立面标准化立方体的堆砌如同数码组合一般简单而富有变化，盒子与盒子之间的抽空形成了趣味性空间，成为建筑内部与外界信息沟通的窗口。

整个大楼光洁、现代、具有极强的可识别性，建筑暗合着数字信号的特征，流动着、延伸着，蔓延至基地周边，并覆盖到技术楼，使得整个场地、新老建筑无论从功能上还是空间上都形成富有变化的有机的整体。建筑形体在自然光的照射下，实与虚、光与影，显示出无穷的变化，绽放出独特的魅力。

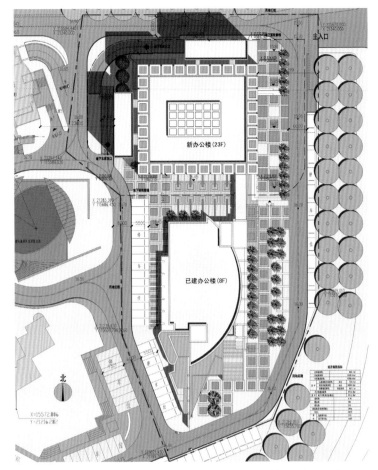

总平面图 Site Plan

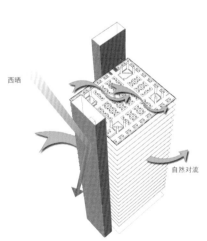

该建筑的设计中解决了常规的高层办公建筑因核心通内置而经常面临的缺少自然通风、交通核无自然采光等问题，并且西侧的核心筒可有效避免西晒的影响，使得建筑有效的减少能源的消耗。

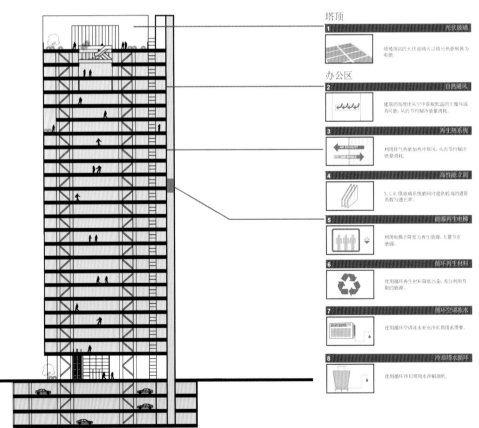

建筑节能
Construction energy conservation

Profile

Shenzhen Media Group Cable TV Hub Building (tentative name) locates at No. B306-0005 plot, the southern Lianhuazhi Road and the western Caitian Road in Futian District, Shenzhen. This site is 2 km away from the centre of downtown Shenzhen, which belongs to the "universal central extension " district at the northward extension of central Shenzhen. The total land area of this project is 8,661.1 m^2 and the design height reaches to 100 m (23 floors). After it is to be finished, it is mainly used for the office purpose for Shenzhen Media Group and Shenzhen Topway Video Communication Co., Ltd.

Planning Layout

This project faces the problem of land shortage and spatial restriction, but the bottom elevated style could effectively ease the oppressiveness. The terrace landscape design balances the height difference between the east and the west, so as to make the space more smooth and insightfully. here people stay and socialize, as a result , this place turns to be an information exchange site. In the meanwhile, it also breaks the barrier between inner space and urban space, strengthens the connection between two urban public landscape —Lotus Hill and Beacon Hill, and their users.

Architectural Design

The piles of cubes with standard facades is simple but full of variations; the evacuation among cases form an interesting space, which becomes the window for the use of communication between the inner building and the information from the outside world.

This whole building is glossy ,modern and posses a strong identifiability. It coincides with the features of digital signals, flowing , extending, until reaching to the base surroundings as well as expanding to the technology building. Therefore, no matter functions or spaces, the whole site and the new or old buildings both develops to be an organic whole with variations. Under the exposure of natural light, the architectural form shows infinite variations ,like solid and void, light and shadow etc. and thus bursts forth its unique glamour.

URBAN INDUSTRIAL COMPLEX

| Shenzhen Skyworth Shiyan Science and Technology Industrial Park Phase II

都市型工业综合体——深圳创维石岩科技园二期

项目地点：中国广东省深圳市
建筑设计：香港华艺设计顾问（深圳）有限公司
总建筑面积：350 000 m²

Location: Shenzhen, Guangdong, China
Architectural Design: Hongkong Huayi Designing Consultants (Shenzhen) Co., Ltd.
Total Floor Area: 350,000 m²

项目概况

项目位于深圳市宝安区石岩镇深圳创维科技工业园。园区总占地411 164.53 m², 其中一期工程已建成使用,二期项目规划分科研、生产、生活三大功能区,总建筑面积350 000 m²。园区集科研办公、生产厂房、宿舍、培训中心以及商业配套于一体。

规划布局与建筑设计

项目打破传统的规划模式,以清晰的规划结构、紧凑的功能布局,打造出高效共享的"都市型工业综合体"。引入"数字绿脉"、"企业绿洲"的设计理念,既有效整合创维工业园一、二期的功能,又能打造"具有生态工业旅游元素的后工业区"环境,塑造出一个更具活力与开放性的新创维形象,同时也树立了深圳新兴生态旅游工业区的典范。

COMMERCIAL BUILDINGS | 商业地产

Profile

The project site is at Shenzhen Skyworth Science and Technology Industrial Park, Bao'an District of Shenzhen. The total land area of the park is 411,164.53 m^2. The first phase of this project has been built to use. The phase II is divided into 3 sections– for scientific research purpose, for production purpose and for living purpose. The total floor area of the park is 350,000 m^2, where gathers scientific and office buildings, manufacturing facilities, dormitories, training centers and commercial support facilities into one.

Layout Planning and Architectural Design

This project breaks the rigid planning mode, instead, it is designed with a clear planning structure and a compact layout. Thus a high efficiency and sharable " urban industrial complex" is created. Apart from that, the imported design concept–"digital greenway" and " companies oasis" ,not only effectively integrates functions of phase I and II of the Skyworth Industrial Park , but also builds up an environment of "a post-industrial park with ecological industrial tourism elements." In the meanwhile, a new image for Skyworth is to be established, which is setting up a model for Shenzhen emerging ecological tourism and industrial district.

COMMERCIAL BUILDINGS | 商业地产

FLEXIBLE AND CONVENIENT, DEVELOPING WITH ADVANTAGES

| Phase II and III of Xiamen Aviation Development Base

灵活便利　应"势"生长——厦门航空综合开发基地二、三期工程

项目地点：福建省厦门市
建筑设计：香港华艺设计顾问（深圳）有限公司
建筑面积：68 164 m²

Location: : Xiamen, Fujian
Architectural Design: Hongkong Huayi Design Consultant (Shenzhen) Co., Ltd
Area: 68,164 m²

项目概况

本项目地处厦门岛东北角的高崎国际机场附近。基地虽不属闹市区，但却是大厦门整体规划的中心区域，交通便利、视野开阔、环境优美，与拥挤喧嚣的市中心区域相比有其独特的地理优势。

规划布局

"围合式的庭院建筑布局"。依靠自身成"势"吸引市场，弥补周围地区开发不成熟所带来的不足。利用围合式布局将二、三期建筑体量进行分解和重组，使其形成一个小型建筑群体。营造出项目自身气氛，充分造势，吸引市场。同时内向型的围合式布局很好的形成内敛、平静、高尚的氛围感。

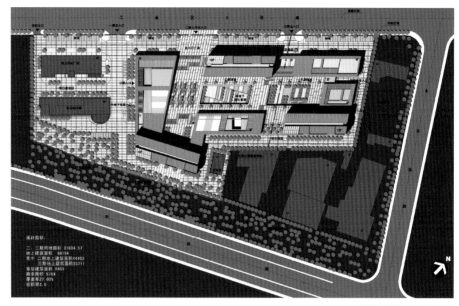

总平面图 Site Plan

建筑设计

"板式主裙楼有机结合"。利用项目容积率较低的优势,将主裙楼都设置为板式体量,有效利用足够面宽,为大、中、小单元的灵活重组提供优质前提。同时将主楼位置相互错开,保证了每栋主楼都有良好的视野,最大化的利用了基地内外景观。而裙楼则更多考虑和内院的接地处理,通过高低、架空、退让的处理方式,营造了一个尺度适宜的建筑群体空间。

"可生长型的分期建设"。方案中人行动线及中心庭院都可以沿一二三期顺序"生长",保证后期建设的协调与融合,同时交通系统也能纳入到统一体系中,共同形成服务于整体项目的流线体系。

"灵活单元划分重组"。作为以租赁性质为主的办公楼建筑,在方案中设置出更为灵活的办公单元面积划分和重组模式。以最经济的垂直交通和水平交通的设置,获得以小面积单元为主,不同面积规模皆备的灵活性办公模式,最大适应多样的市场要求。

"经济价值性能"。方案通过对二期设置半地下停车、三期设置架空停车和地下停车的方式,一方面有效缩减车库的建造费用,另一方面以一个低成本方式为基地内提供了一个立体园林。从总体布局设置与单体灵活设置两方面使建筑物对灵活多变的市场需求保持足够的敏感性及适应灵活性,以确保项目在未来的市场竞争中具备足够的潜在竞争优势。

Profile

This project is close to Gaoqi International Airport which locates in the northeast corner of Xiamen Island. Though far away from the downtown area, it is the center area in the overall planning of Xiamen City. It enjoys convenient traffic, open views and beautiful environment which are superior to the busy and noisy downtown.

Planning and Layout

Enclosed courtyard layout. It takes advantages of itself to draw the attention of the market. Buildings of phase II and III are decomposed and re-organized in courtyard layout to form a small building complex, which will shape its own identity to attract customers. At the same time, this kind of layout will provide a peaceful and high-end environment.

Architectural Design

Slab-type podiums are well arranged. It takes advantage of the low plot ratio to design the podiums in slab-type which will provide flexibility for the re-organization of units in different sizes. Meanwhile, tower buildings are staggered to ensure great views of the landscapes, while podiums are well designed to connect with the inner courtyard and provide a comfortable space.

Growing and Developing Project. The traffic system and the central courtyard are developed stage by stage to ensure the coordination and integration of the whole project.

Flexible Separation and Recombination of Units. As a office building mainly for rent, the design provide high flexibility for the separation and recombination of units in different sizes. Offices of different sizes (mainly small-sized ones) are set around the economical vertical and horizontal traffic system to meet different requirements which will help it to win the market in the future.

Economic Value. The plan greatly reduces the construction cost for garages by setting semi-underground parking in phase II as well as elevated parking and underground parking in phase III. In this way it also provide a three-dimensional landscape system for this base. From both the overall layout to the single building design, it pays attention to the flexibility of the spaces to meet different requirements.

ATTRACTIVE URBAN SPACE: OPEN AND NATURAL

| Super High-rise Complex of Global Valley in Zone C of The Future Ark, Guiyang

开敞自然 极具吸引力的城市空间
—— 贵阳未来方舟C区环球谷超高层综合体

项目地点：贵州省贵阳市
建筑设计：香港华艺设计顾问（深圳）有限公司
总建筑面积：266 000 m²

Location: Guiyang, Guizhou
Architectural Design: Huayi Design & Consultant (Shenzhen) Co., Ltd.
Total Floor Area: 266,000 m²

项目概况

环球谷项目位于贵阳市云岩区，是中天未来方舟区域开发项目的子项目之一。规划定位为未来方舟项目四大门户之一，兼具标志性和区域配套服务功能。环球谷由一栋320m高的超高层塔楼、一栋100m的商务公寓和一个大型山地购物中心组成。塔楼部分包括白金五星级酒店和高端商务写字楼。

建筑设计

超高层塔楼如新生的竹笋破土而出，建筑的弧形表皮充满动感；商务公寓采用与超高层相同的设计手法，并用低矮的体量衬托出塔楼的高大；山地购物中心引入了体验式休闲商业的概念，以开放的体验化空间吸引人们主动游玩和消费。项目颠覆了一味追求机械效率的工业化空间设计，打破了室内室外的空间界限，实现了城市森林中自然化、戏剧化空间场景的塑造，实现了为人们提供聚会、游玩、欢庆场所的目的。

COMMERCIAL BUILDINGS | 商业地产

Profile

Located in Yunyan District of Guiyang City, Global Valley is part of The Future Ark project which will be one of the four gateways serving as the landmark and supporting facility. It is composed of a 320m towering building, a 100m service-apartment building and a large-scale shopping mall. The towering building will house a platinum five-star hotel and the high-end offices.

Architectural Design

The towering building breaks through the soil just like the bamboo shoot. Arc-shaped skin is full of dynamic sense. The apartment building is designed lower in the same style, to highlight the height of the tower. The shopping mall introduces new leisure and commercial concept, trying to attract people with open and experience space for enjoyment and shopping. The project gets rid of the industrialized space design, breaks the boundary between indoor and outdoor spaces, creates natural and dramatic spaces in the urban forest, and provides an ideal place for gathering, enjoyment and celebrities.

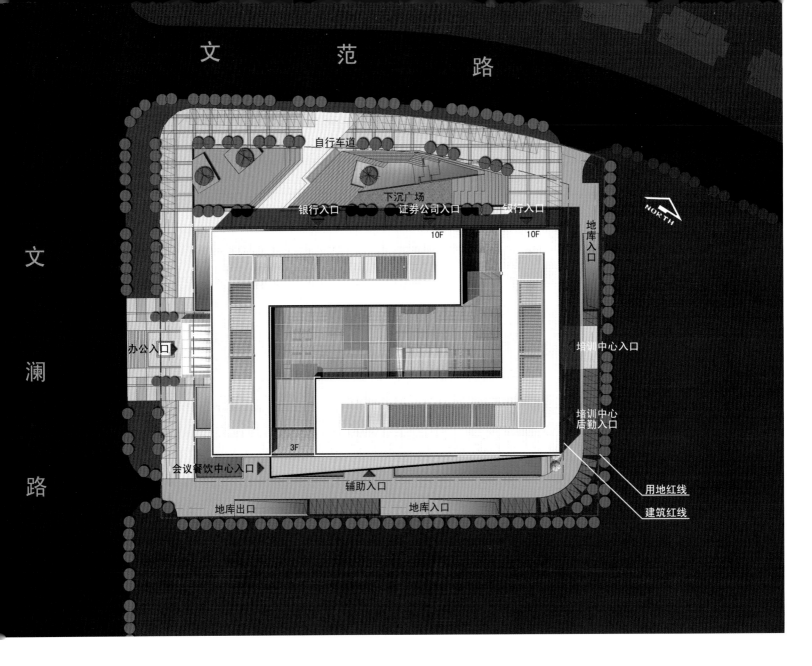

MODERN MINIMALIST AESTHETICS AND TECHNOLOGY COMBINED HEADQUARTERS | NANJING CSCEC Grand Tower

美学与技术相结合的现代简约总部形象 —— 南京中建大厦

项目地点：中国江苏省南京市
建筑设计：香港华艺设计顾问（深圳）有限公司

Location: Nanjing, Jiangsu, China
Architectural Design: Hong Kong Huayi Design Consultant (Shenzhen) Co., Ltd.

项目概况

项目位于南京市栖霞区仙林大学城区。总用地面积21 778.92 m², 计容积率总建筑面积约60 000 m²。地上11层, 地下1层, 高度为49.8 m。

建筑设计

建筑造型采用整体形式, 取意为"玉筑", 独特的建筑形象将为中建公司树立特色鲜明的总部形象, 更将成为大学城中心区的重要地标建筑。"玉"和"筑"也分别代表了美学与技术相结合的设计逻辑。建成后的中建大厦将集办公、商务十一体, 既融合建筑现代简约的高档品味, 又充分融合中建安装公司大型国有企业形象。

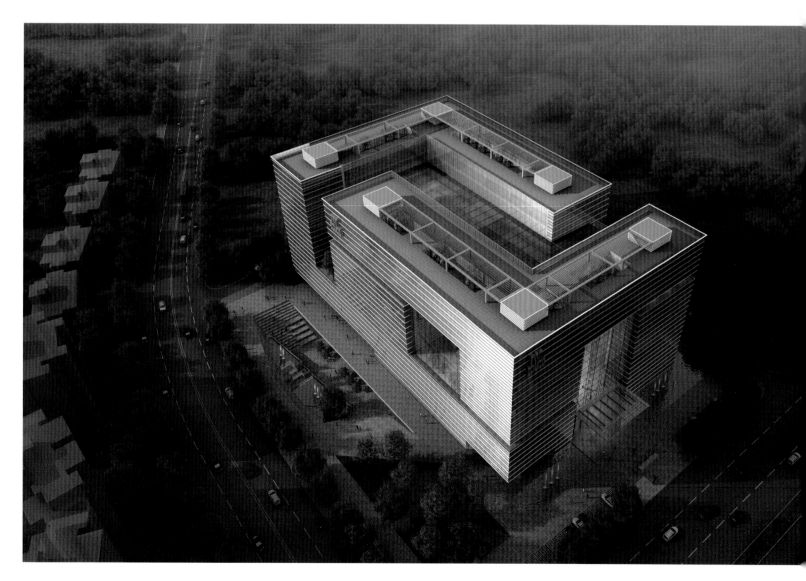

COMMERCIAL BUILDINGS | 商业地产

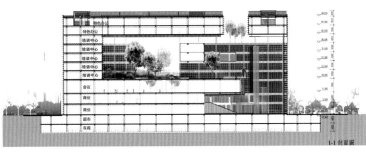

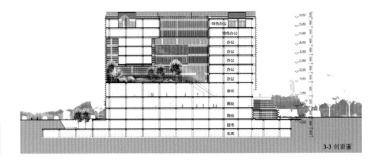

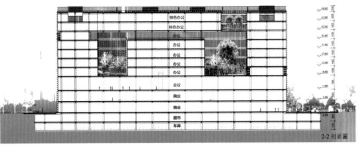

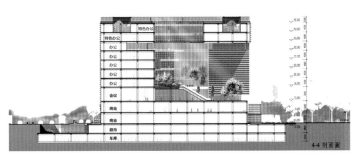

Profile

Located in Xianlin University Town in Qixia District of Nanjing City, occupying a total land area of 21,778.92 m^2, the building has a total floor area of 60,000 m^2. With eleven floors on ground and one floor under ground, the total height is 49.8 m.

Architectural Design

The building form is integrated to imply "jade construction". Unique appearance will highlight the characteristics of Zhongjian Corporation. It will become an important landmark in this area. "Jade" and "Construction" symbolize the combination of aesthetics and technology. Upon completion, CSCEC Grand Tower will be a modern office and business building with high-end taste to show the identity of its owner.

COMMERCIAL BUILDINGS | 商业地产

LYRE PLATFORM AND NIGHT MOON | Shanxian Culture and Art Center

琴台夜月 —— 单县文化艺术中心

项目地点：中国山东省单县
建筑设计：上海中建建筑设计院有限公司
总建筑面积：50 000 m²

Location: Shanxian, Shandong, China
Architectural Design: Shanghai Zhongjian Architectural Design Institute Co., Ltd.
Total Floor Area: 50,000 m²

项目概况

单县文化艺术中心位于山东省单县君子路以东,单州路以南,处于单县新城中心。项目西邻玉泽河,东邻东沟河,南邻玉泽湖,三面环水,风景优美,北部为大型商贸广场、湖西人民广场和新政府大楼,地理位置优越。用地面积约86 000 m²,文化艺术中心地上建筑面积约45 000 m²,地下建筑面积约5 000 m²,总建筑面积约50 000 m²,项目同时还包括一个5 000 m²配套酒店。

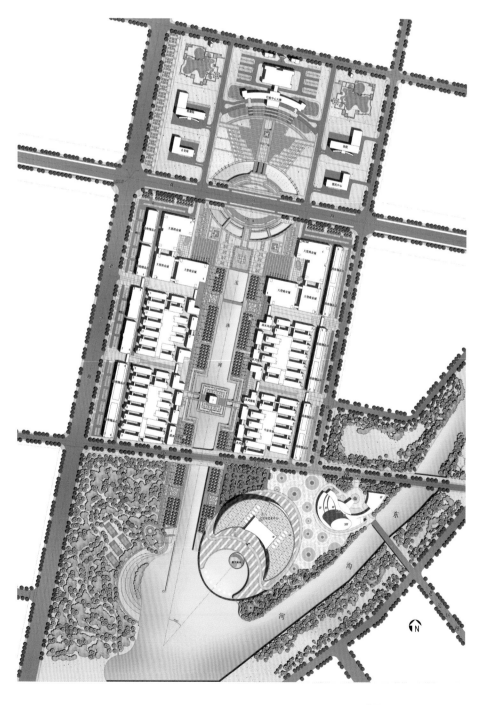

总平面图 Site Plan

COMMERCIAL BUILDINGS | 商业地产

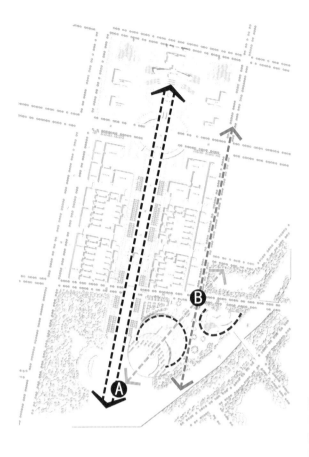

- Ⓐ 规划轴线交点
- Ⓑ 视觉轴线交点
- 空间主轴线
- 建筑轴线
- 视线通廊
- 建筑界面

规划布局

项目规划主要用于文化、商业、展览以及休闲等，主要功能分区包括演艺中心、活动中心、以及文化中心，同时室外还有艺术文化广场、露天舞台、市民广场、历史文化广场等。文化艺术中心的基地位于行政中心轴线的尽端，临水而筑，这是一处巨大的开敞空间，基地周围已建有一些人们所关注的道路，人行道和湖滨公园，所以文化艺术中心与行政中心、公园及水系的种种联系显得密不可分。这些已落成建筑的尺度和环境会绝对地影响文化艺术中心的布局，使其成为整体空间布局里的一个重要的节点，一个文化的焦点，一个人气的地标。

建筑设计

单县古琴台台高数丈，呈半月形，每当皓月当空，登台远眺，银辉洒处，朦胧如水，心旷神怡，如置身仙境。脍炙人口的故事总是揉合着这诗一般的美景，"琴台夜月"广为流传。相传春秋时单父宰宓子贱曾在此抚琴，留下"鸣琴而治"的佳话。以"琴台夜月"为主题的设计理念源于对单县历史

文脉的传承，也是项目的特色和亮点。整体造型新颖、独特、美观，是一张独一无二的、单县特有的文化名片。

设计师期望建筑能在绿荫环绕的玉泽河尽端突然出现，宛如一座神秘、和谐、而又具有活力的雕塑从水边的树林中忽现。无论从哪个视角，包括从周围高楼上的俯瞰，它都应该是一座能为人们展示不断变幻而吸引人的雕塑，正如其它标志性建筑一样，使人产生强烈的印象，并赋予以个性。文化艺术中心的幕墙使用一种因时而异的表皮肌理。白天，她披着神秘的面纱，纯净的白色在阳光下熠熠生辉；夜晚，晶莹剔透，流光异彩，尽情地向人们展示其美妙的内部空间。

COMMERCIAL BUILDINGS | 商业地产

Profile

Shanxian Culture and Art Center is located in center of Shanxian New Town, the east of Junzi Road and the south of Danzhou Road. It occupies a superior geographic position that adjoining Donggou River in the east, Yuze River in the west, large-scale business trade center, Huxi People Plaza and New Government Building in the north. It has a total land area of 86,000 m^2, of which the ground floor area is 45,000 m^2, underground area is 5,000 m^2, and the total floor area 50,000 m^2, including a 5,000 m^2 supporting hotel.

Planning and Layout

The project is mainly functioned in culture, commerce, exhibition, and leisure, etc. The main functional zones are art performance center, activity center and culture center. Besides, there are art and culture square, open stage, civic square, historical and cultural square, etc. The construction base of culture and art center is situated at the end of the administrative central axis, built along the river. It is a huge open place with remarkable roads, sidewalk and Lakeside Park, so the culture and art center gets a tight relationship with the administrative center, park and the river system. The completed buildings and surroundings are certainly to make a great influence on the layout of the project and make it to become an important part, a culture focus and a landmark in the whole layout.

Architectural Design

The lyre platform in Shanxian is more than ten meters high, looking like a half moon. When the bright moon rises in the sky and someone looks far into the distance on the platform, silver moonlight sprinkles hazily that makes people relaxed and happy, just like in the fairyland. Thrilling story always goes together with the poetic beauty, so "lyre platform and night moon" is widely circulated. According to the legend in the Spring and Autumn Period, a man called Mi Zijian had played the lyre in this platform, so "lyre platform and night moon" emerged.

The design concept of "lyre platform and night moon" derives from the inheritance of ancient culture of Shanxian, which is also the highlight of this project. The whole design is novel, special, and beautiful, being the unique cultural characteristic of Shanxian. The project breaks out in the end of the Yuze River like a mystical vigorous sculpture popping up in the forest. Seen from different angles, it just looks like changing and attractive sculpture, leaving people a strong impression like other symbol buildings. The curtain walls of the culture and art center apply a time-changing surface texture. In the daytime, it wears a mystical veil, shining in the clear sunlight. At night, it looks crystal clear and extraordinary, showing people its splendid inner space.

www.cihaf.cn　010-65079988

推动中国房地产向上的力量

中国地产节
——房地产产业链全景展示交流平台

2012/12/6-12/8　深圳会展中心

500位行业领袖及嘉宾

13年来

3座举办地

60000名专业观众

1200家供应商

CIHAF 2012 第十四届中国（深圳）国际房地产与建筑科技展览会
The 14th China International Real Estate & Architectural Technology Fair

城市中国　地产中国　设计中国　绿色中国　家居中国　文旅中国　服务中国　中国之家　CIHAF论坛
URBANIZED CHINA　PROPERTY CHINA　DESIGN CHINA　BUILDING CHINA　FURNISHING CHINA　CULTURE TOURISM　SERVICING ESTATE　CHINA HOME　CONFERENCE

www.bacdesign.com.cn

住宅景觀・私家庭院
Residential Landscape・Private garden

酒店景觀・城市公共景觀
Hotel Landscape・Urban Planning Design

旅游度假項目規劃
Resorts and Leisure Planning

建築景觀模型
Architecture model

長期誠聘景觀設計人才,誠邀專業人士加盟.
地址:廣州市天河區廣園東路2191號時代新世界中心南塔2704室(郵編:510500)
Add:Rm., 2704, South Tower, The Times New World Center, No., 2191, Guangyuandong Rd., Tianhe Dist., Guangzhou(P.C. 510500)
電話Tel:020-87569202　(0)13688860979　Email:bacdesign@126.com

四季园林

鸿艺集团.客天下.瀑布教堂

- ■ 风景园林专项设计乙级
- □ 景观设计　　　　Landscape Design
- □ 旅游建筑设计　　Tour Architectural Design
- □ 旅游度假区规划　Resorts and Leisure Planning
- □ 市政公园规划　　Park and Green Space Planning

广州市四季园林设计工程有限公司成立于2002年，公司由创始初期从事景观设计，已发展为旅游区规划、度假区规划、度假酒店、旅游建筑、市政公园规划等多类型设计的综合性景观公司。设计与实践相结合，形成了专业的团队和服务机构，诚邀各专业人士加盟。

Add：　广州市天河区龙怡路117号银汇大厦2505
Tell：　020-38273170　　　　Fax：　020-86682658
E-mail:yuangreen@163.com　　Http://WWW.gzsiji.com